LEAP
OF
FAITH

LEAP
OF
FAITH

An Astronaut's Journey into the Unknown

GORDON
COOPER

with Bruce Henderson

HarperCollins*Publishers*

For legal reasons, some of the names in this book have been changed.

FIRST EDITION

Designed by Nancy B. Field

Library of Congress Cataloging-in-Publication Data

Cooper, Gordon.
    Leap of faith : an astronaut's journey into the unknown / Gordon Cooper, with Bruce Henderson — 1st ed.
        p. cm.
    Includes index.
    ISBN 0-06-019416-2 (alk. paper)
    1. Cooper, Gordon. 2. Astronauts—United States—Biography. 3. Life on other planets. I. Henderson, Bruce. II. Title

TL789.85.C66 A3 2000
629.45'0092—dc21
    [B]   99-086433

00 01 02 03 04 ❖/RRD 10 9 8 7 6 5 4 3 2 1

For my wife, Suzan,
who has put up with a lot through the years
from this scruffy old fighter pilot

LEAP
OF
FAITH

PROLOGUE

They woke me up at three in the morning, in the dark.

I had been up late the night before, hitting the sack not much before eleven o'clock in Hangar S, the astronaut head-quarters at the Cape. The hangar had been remodeled inside to accommodate our various types of training apparatus, a pressure chamber, living quarters and a dining room, a medical suite, a ready room, and everything else we needed to prepare for a mission. It was also here that the National Aeronautics and Space Administration (NASA) had offices and where our spacecraft were constantly worked on by engineers and technicians.

I showered, then enjoyed a "going away" breakfast of fresh-squeezed orange juice, filet mignon, scrambled eggs, toast, grape jelly, and coffee, with some of the key people on the team, including the flight surgeon and a few fellow astronauts who, through their good-natured ribbing, were determined to see that I stayed loose.

From there, I went down the hall to the suiting-up room—a sterile space off-limits to bugs, dirt, debris, and anything else we didn't want to take along into space. Everyone working in here was dressed in a white smock and white hat and wore disposable paper covers over their footwear.

I stripped, and a medical technician glued a half-dozen medical sensors to various spots on my body that he'd first sandpapered and scrubbed with alcohol. I donned long underwear and was helped into my bulky flight pressure suit.

I took the seven-mile trip to pad 14 in the back of a NASA van.

I much preferred the way I had arrived at the pad for final "hot tests" of all systems several days earlier: at the wheel of my new 300-horsepower blue Corvette, which I had purchased—the same way other astronauts purchased their 'Vets—at a 50 percent discount from General Motors through its Brass Hat program for employees and a limited number of VIPs. That day I had successfully lobbied my NASA bosses to let me drive myself to the pad on the promise that I'd be careful. I hadn't even bothered to ask this morning, knowing what the answer would be.

My rocket awaited me: Atlas booster number 130D, more than ten stories tall. It stood in the pitch dark like a lone sentry, eerily bathed in columns of white light thrown skyward by huge spotlights.

The first time I had seen an Atlas launch was three years earlier, only a few months after seven of us had been selected for Project Mercury as America's first astronauts. They took us down to the Cape for a launch. We stood as a group outside in the humid night—the "Mercury Seven," as we became known—flat-topped fighter jocks in short-sleeved shirts, buoyant and more than a tad cocky, as all good fighter pilots must be. We came to be referred to on in-house memos as CCGGSSS, as our names were always listed in alphabetical order: Carpenter, Cooper, Glenn, Grissom, Schirra, Shepard, Slayton.

We were there to see a test of the Atlas, the newly built

rocket that was expected to carry us into orbit in the years to come. We watched the huge rocket rise slowly off the pad on three thick columns of flame. It was a thunderous start, and the ground rumbled beneath our feet. The rocket flew straight up for forty or fifty seconds and was almost directly overhead as it slowly began to nose over on its long arcing flight toward the horizon. Several hundred VIPs were in attendance, including many congressmen and several committee chairmen who were instrumental in providing funding for the space program. We all stood there, craning our necks to keep our eyes on the magnificent silver bird, when suddenly the sky erupted in a tremendous *kaboom!* and the rocket blew up in a jillion flaming pieces. While some spectators ducked instinctively, the burning debris would be carried by the rocket's momentum out over the ocean and drop into the drink several miles offshore.

We looked at each other. "That's our ride?" someone deadpanned.

Two months later there had been another cataclysmic failure, and in the months that followed, more failed attempts.

In late 1960 NASA put on another public show designed to prove that the Mercury spacecraft and the smaller Redstone rocket, which would be used for two planned suborbital flights, were ready for upcoming manned missions. These flights would not go high or fast enough to reach orbit but would simply slingshot the spacecraft one hundred miles high, where space officially begins—some fifty miles above the Earth's atmosphere—and downrange three hundred miles at five thousand miles per hour for an ocean landing near Bermuda. For the event, NASA brought the seven of us back, along with hundreds of VIPs. This time, as the flames burst from the rocket, it lifted off only two inches before the engines mysteriously cut off and the rocket settled back onto the pad. All was quiet. However, the signal for liftoff had already been sent to the automatic sequencer. About sixty seconds later—as called for by the flight plan—the spacecraft's escape tower, designed to pull the capsule and its occupant to safety in the event of a problem in the first minute after liftoff, separated from the spacecraft and shot into the sky like a spear. As the crowd silently watched the only thing moving out there that day, the escape tower went up about four thousand feet, then came back down, driving itself into the ground about a hundred yards away.

As for the Atlas, before John Glenn's orbital mission in 1962—the first manned mission to use the powerful new rocket—there had been a total of thirteen blowups, some on the launch pad and others as the Atlas climbed into the sky. Test pilots understood the nature of new technologies: going up in hurtling pieces of untested machinery and putting our hides on the line. We knew about hanging it over the edge and pushing back the envelope, then hauling it back in and doing it again tomorrow. When the time came, each of us would take his turn sitting atop the same type of rocket we'd seen blown to smithereens before our eyes. We did so not because we were suicidal or crazed but because we were pilots.

Early on in the space program there had been quite a debate over whether the manned launches should be covered on live television or a taped and edited version offered to the public later. Live TV would reveal anything and everything that went wrong; in the event of a tragic failure, public sentiment could turn against the program. I suppose we could have tried to do what the Soviets did—hide our failures and trumpet our successes—but could a free country really do that? We were the representatives of an open society, and we would take our chances in full view of the world—on live television.

Up the gantry elevator I went on launch morning, carrying the suitcaselike portable cooler that kept me supplied with oxygen until I could be connected to the spacecraft's system. At the top I was greeted by Guenter Wendt, who in his "clean room" attire of white overalls and white sandwich hat looked like the neighborhood ice cream man but who was one of the brilliant German rocket scientists helping the United States in the space race with the Russians, who also had German rocket men assisting them.

I saluted Wendt, keeping up an old private joke. "Private Fifth Class Cooper reporting for duty."

He saluted back. "Private Fifth Class Wendt standing by."

We had been "demoted" following a practical joke pulled a couple of years earlier on a TV film crew. We had all been working around the clock getting ready for the first suborbital mission when, about three days before launch, the bosses decided to let a news crew film "launch day in the life of an astronaut." I was selected to be the model, and they followed me everywhere, even filming me getting into long underwear after the sensors were attached to various parts of my anatomy. All of us considered the contrived show a big waste of time, but NASA was always looking to curry favor with the media. We had gone out to the pad in the transfer van, and when we got there Guenter met us. As we arrived at the bottom of the elevator, the door opened and everyone started forward. I suddenly grabbed the door of the elevator and hollered, "No! No! I won't go!" Guenter, meanwhile, was trying desperately to drag me into the elevator. The NASA public relations people and the media were not amused. One writer suggested that Guenter and I should be busted to "Private Fifth Class," and it stuck.

As I sat atop the rocket, at some point during the final hour of the nearly three-hour preflight routine, after I had done all my work and we were waiting for technicians handling radar and other tracking devices to do their final calibrations and get up to speed, I felt myself begin to nod off. They had gotten me up *so* early. Knowing how busy I would soon be, it seemed like an ideal opportunity to grab a nap.

So, sitting atop the fully fueled rocket, with no place to go for the time being, I fell fast asleep.

"Gordo!"

I came awake instantly, immediately aware of my situation. "Uh-huh."

"Hate to disturb you, old buddy," said Wally Schirra, serving as capsule communicator (CapCom) for my launch, "but we've got a launch to do here."

"Let's do it. Ready when you are."

As the last of the Mercury astronauts to get a space shot—I like to think they were saving me all along for the last, the longest and best mission—I *was* ready . . .

WE SEVEN

In early January 1959 I received unexpected orders to report to Washington, D.C. At the time I was a thirty-two-year-old air force captain with twelve years of service, based at top-secret Edwards Air Force Base in the California desert.

The best test pilots in the air force were assembled at Edwards, and I had a great job. As a test pilot in the engineering branch, I had the best of both worlds. I saw a project from the design and administrative side as well as from the pilot's seat. I was testing and flying the nation's top new aircraft—hot fighters like the F–102 and F–106, and the secret U–2, a reconnaissance aircraft built like a graceful glider (and not much faster than one) with an unusually long wingspan and lightweight fuselage, that could fly higher than the surface-to-air missiles of the day.

A day or two before I was due to leave for Washington, I was called into the base commander's office with three other test pilots—including one named Donald "Deke" Slayton—who had received similar orders.

Our commanding officer, General Marcus F. Cooper (no relation), asked if any of us knew what our orders were about.

"No, sir," we all replied.

"No one will tell me anything," the general groused.

He was a good CO, a lot better than the old-stick-in-the-mud general he'd replaced a year earlier. General Cooper remembered what it was like to be a young pilot, and he was very protective of his men as long as we did the job he asked of us.

"I did see in the paper the other day," he went on, "that McDonnell Aircraft has been awarded a contract for the new manned space program that's starting up."

My ears perked up. I knew nothing about any "manned space program."

"Gentlemen," General Cooper said with authority, "if this has anything to do with flying in space, I want you to be *very careful* what you volunteer for. I don't want my best pilots to be involved in some idiotic program."

It had been a little over a year since that historic day in October 1957 when the Russians launched Sputnik, the world's first manmade satellite. The 184-pound satellite—about the size of a basketball—could be heard by American tracking stations as it orbited Earth making a characteristic "beep-beep" sound. Residents of neighborhoods across the country waited anxiously in their yards and streets, peering into the sky at a fast-moving speck of light that was surprisingly easy to see.

I had realized that Sputnik stood to open up a whole new era, and that the Soviet Union had the potential for being one up on us militarily. It reasoned that people and events on Earth could one day be observed from space. When that happened, there would be nowhere to hide.

Two months later, the U.S. Navy attempted to launch the first American satellite. It was also the first nationally televised rocket launch. Upon completion of the countdown, the Vanguard rocket lifted less than a foot off the ground before the first stage, loaded to the gills with fuel, exploded. The rest of the rocket sank in slow motion toward the ground, embedding itself in the sand next to the launch platform like a burned-out firecracker. It left an indelible image of America's opening bid in the space race.

Were they now actually thinking of strapping a man onto a rocket?

Unbeknownst to any of us, qualifications for prospective astronaut-pilots had been established by NASA, the newly founded civilian agency designated to lead America's efforts in space, which had received funding from Congress only after Sputnik. Although the United States did not yet have the spacecraft and other hardware necessary to send a man into space, NASA came up with a list of specific requirements to describe the kind of space pilots it was seeking.

It was believed that they needed to be in their physical prime while possessing a degree of maturity to handle difficult situations. Maximum age was set at forty. Maximum height was five feet, eleven inches, an arbitrary cutoff that eliminated any number of otherwise well-qualified pilots. The spacecraft already on the drawing board had its dimensions dictated by the diameter of the available boosters—the Redstone and Atlas missiles—that would launch it into space. Those dimensions were seventy-four inches wide at the base, and it was determined that once a pilot had on helmet and pressure suit and was strapped in for liftoff, anyone six feet or taller wouldn't be able to squeeze inside.

The weight limit was set at 180 pounds for two reasons. The first had to do with the finite (and limited) payload of the available boosters—the more a man weighed, the less room there would be for the equipment that would be necessary for a safe and successful mission. Just as important was the belief that anyone who met the height requirement but weighed more than 180 pounds would probably be overweight and therefore have a less than optimal metabolic and circulatory system to handle the stresses of prolonged weightlessness and rapid changes in temperature.

The search for candidates narrowed to the ranks of practicing test pilots—meaning pilots from the air force, navy, and marines, along with a handful of civilians. The theory was that test pilots had the instincts and training required to handle a complex spacecraft traveling at high speeds and high altitudes. It made sense that the same men who were testing our country's hottest jets should be in the driver's seat when it came time to launch a manned space vehicle.

A thorough search of personnel records came up with the names of 508 test pilots who met the basic requirements. That list was reviewed further and whittled down to 110. Then a special NASA committee on life sciences, partly on the basis of confidential evaluations of candidates supplied by instructors who had taught the pilots how to fly and others who knew the quality of their nerves and reflexes, narrowed the list down further: to 69 prospective candidates, who were ordered to Washington, D.C.

We met on February 2, 1959, in a large briefing room at NASA headquarters in downtown Washington, not far from the White House. NASA administrators and engineers spent the entire morning giving us a rundown on the space program and what part the astronauts would play.

Project Mercury, the free world's first program for the manned exploration of space, was so named, we learned, for its symbolic meaning: Mercury was the winged messenger of Roman mythology. The Mercury mission was designed to explore and develop technology necessary to launch a man into orbit. We were told by a group of enthusiastic NASA officials that this would involve newly developed techniques in aerodynamics, rocket propulsion, celestial mechanics, aerospace medicine, and electronics.

"Gentlemen, you have an opportunity few men have even dreamed of. . . ."

The sales pitch was on. When they got to the part about launching chimpanzees first, there were some raised eyebrows among the fighter jocks. But we knew that if we wanted to go higher and faster and farther, as we all did, this was the way to go. And when I saw what a logical step-by-step program NASA had in mind, and what a major role the astronauts would have in it—not only as pilots but in the engineering development of the program as well—I was pretty sure I wanted in. My only doubts involved having to leave Edwards and all the fun I was having there, for a new civilian start-up program that might or might not make it.

I had been flying some real high-performance airplanes and was getting up to what we then considered high altitudes and high speeds. I had the natural desire of most pilots to go higher and faster. As for space, I had long thought we ought to be trying to extend man's capacities up there. But to have it presented as a real program, not just some Buck Rogers fiction, was quite a jolt.

Would it actually come to pass? I wondered.

I considered what it would be like to be strapped onto the top of a big rocket and blasted off into the dark realms of space. If I volunteered and ended up being chosen for the program, would I be able to overcome the fear of the unknown and do a credible job of flying the vehicle? These were the same questions I asked myself on a regular basis as a test pilot. The real fear, for me, had always been the uncertainty of the unknown. Would I find a big surprise that I wasn't prepared for? And if so, would I come up with a way of handling that surprise in order to preserve my life and the mission?

At the end of the day, we were given the choice to return to our bases, no questions asked, or volunteer for continued testing. Thirty-seven men, most of them not willing to make such a radical change in their careers, dropped out of the running and returned home. Thirty-two of us volunteered to take a chance and move on to what promised to be an exhaustive series of medical examinations and psychological tests.

I knew some of the other pilots who were being considered, guys like Deke Slayton from Fighter Operations at Edwards, for example. Some of the navy guys seemed sharp too. I could see that there was going to be stiff competition for what we were told would be a dozen astronaut jobs. At that point I felt I would be lucky if I made the grade, but I wanted a chance to fly in space so much that I was determined to do my best.

The psychological tests, which came first, took many long hours and were hard work but kind of fun too. They were the sort of exams that left you not knowing how well you had done, or even if there were right or wrong answers. The psychologists seemed to be trying to measure our maturity, alertness, and judgment.

There were more than five hundred questions on the personality inventory, designed to dig under the surface in an effort to see what kind of people we were. What was our true motivation for joining the program? Were we too egocentric to work in a team setting? In what the doctors called the "Who am I?" drill, we were asked to complete the sentence "I am ..." *twenty times*. The first five or six were pretty easy—"I am a man," "I am a pilot," "I am a father." But before long, you had to do some serious thinking about exactly who you were—and then hope that the psychologists liked what you found.

We were divided into smaller groups for further testing, and the luck of the draw placed me in the first group of six candidates to move on in the competition. My group included a navy lieutenant commander named Al Shepard, who had been one of the navy's top test pilots at Patuxent River, where the navy tested its hottest new aircraft.

Next we were flown to Albuquerque, New Mexico, for strenuous physical examinations at the Lovelace Clinic, a newly opened private diagnostic clinic that would become the Mayo

Clinic for space-related medical research. At this point, though, much of what they were doing was pure guesswork. The doctors got real creative, coming up with some unusual tests—such as blindfolding you and sticking a hose in your ear and pumping cold water into your ear canal. Just when you thought your eye-balls were going to float away, they would take out the hose and remove the blindfold and jot down some notes on a pad. Any questions like "What's that for?" were met by grunts or some equally dismissive response, as in *Don't worry, you don't need to know*.

Like a bunch of lab rats, we were probed, poked, sampled, tested, and in general completely humiliated on a regular basis for the better part of a week. At one point they had us carrying a gallon jug around everywhere we went to collect a twenty-four-hour urine sample; it wasn't long before the jars began to get real heavy. All the while, the doctors kept finding places to examine us we didn't even know we had.

When it came out during the exams that I was bothered by hay fever, I found myself having to convince the doctors that my allergies wouldn't pose a problem. "I can't imagine I'll run into many mixed grasses and sycamore trees in space," I said.

Next came the physiological stress tests, at Wright-Patterson Air Force Base in Ohio. The doctors there—sadists to a man—knew how to separate the tigers from the pussycats. Over the course of a week, we were isolated, vibrated, whirled, heated, frozen, and fatigued.

No one had any idea of the physical demands that space travel would place on the human body—some "experts" weren't even sure if a man could survive the rigors of launch, and if he did, whether he would be able to swallow under zero gravity and so be able to accept fluids or food. So they did to us just about everything they could think of. And with the large pool of volunteers, they were free to be deliberately brutal, as I think they were, in an attempt to weed out the candidates.

They put each of us into a tub of ice and water for an hour. As I settled in, I sighed, "Oh boy, this is just like trout fishing back home in the mountains." A technician nearby raised an eyebrow and made a notation on his pad that I liked it. When I went into an ovenlike room to be baked at 160 degrees for an hour, I purred, "Oh my, I'm back on the desert. This is the temperature we fly in all the time at Edwards." Another white coat noted that I liked the heat. When they put me in a tiny dark box to test for claustrophobia and fear of isolation, I fell asleep.

They had us run in a chamber simulating high altitude while wearing a partial pressure suit, and a number of volunteers dropped out of the competition at this stage. I had an advantage over some of the others because I had experience wearing one of these contraptions at Edwards. By regulation they were worn by pilots who expected to go above fifty thousand feet—which included the aircraft I flew, like the U–2, F–102, and F–106—in case there was an emergency cabin depressurization. They were designed to keep the pilot alive—breathing, with proper blood flow through the body—until he was able to get down to a lower altitude. The old partial suits—forerunners of the more pilot-friendly full pressure suits that were developed for space flights—had capstans, or tubes filled with air, at different places along the legs, arms, and torso. As you went higher, the capstans would inflate, grabbing and pinching chunks of skin and hair as they did.

In the altitude chamber, which we dubbed the torture chamber, they ran us up to the equivalent of a hundred thousand feet. In a partial suit, breathing was the opposite of what you were accustomed to. To inhale, you relaxed and your lungs would fill up effortlessly. Exhaling was another matter. The only way to empty your lungs was to forcefully exhale with all your might. The process got exhausting pretty quickly.

By the time I returned to Edwards, I was confident I had made the grade. Even though I'd heard that the field of candidates was going to be narrowed and they would be picking only six or seven astronauts rather than a dozen, I suggested to my immediate supervisor that he ought to start looking for my replacement because I was about to be chosen as an astronaut. I knew about how long it would take for the remaining candidates to be tested, and I was not surprised when the phone call came in early April. In fact, at the first ring I had a strange feeling this was the call I'd been waiting for.

"I'm ready," I said, not waiting to hear the caller's identity.

It turned out to be the assistant manager for Project Mercury, Charles Donlon, whom I had met in Washington.

He chuckled. "You still interested in being an astronaut?"

"Yes, sir, I am."

He wanted to know when I could leave for Langley Air Force Base in Virginia, where Project Mercury had its headquarters.

"How about right now?" I said.

"Next Monday will be fine."

One other pilot from Edwards, Deke Slayton, had also made the grade.

There was an additional unspoken requirement for selection: NASA wanted only happily married family men as America's first astronauts. The main reason was a public relations one, although there was also a theory that marital unhappiness could lead to a pilot making a wrong decision that might cost lives—his own and others.

This happily-ever-after business posed a problem for me, because my wife, Trudy, and I had been separated and living apart for some time. I had been living in bachelor quarters at Edwards, and she was in San Diego with our two daughters, then eleven and nine. We had discussed getting back together for the kids but had made no decision.

When the topic of my twelve-year marriage came up during the interview process, I pretended everything was fine and dandy. Yep, couldn't be better. I knew that a phone call or two could put an end to my charade, so as soon as I had a chance I made a quick trip to San Diego to talk to Trudy.

I told her about my opportunity to be selected as an astronaut, explaining that I would retain my military rank and pay while on special assignment to the new civilian space agency. We discussed what the space program might mean for the astronauts and families alike. Trudy, a licensed pilot herself, caught my enthusiasm. We agreed that it could be an adventure for us all and that a reconciliation was called for in order for me to remain a viable candidate for the program. We were soon back living together under one roof.

As far as NASA was concerned, we had never been separated and had an all-American marriage. Trudy and I would both struggle with the "happily married" illusion through the years. Eventually I spent most of my time away from home except when I scheduled things to do with the children. (Other astronauts had similar problems; only three Mercury astronauts were to remain in first marriages.)

The "Mercury Seven" were introduced at a Washington, D.C., press conference on April 9, 1959. Three were from the air force, three from the navy, and one from the Marine Corps. It had been only three months since sixty-nine of us first gathered in a NASA briefing room to learn about the new manned space program.

Our coming-out party was no small affair—the largest auditorium at NASA headquarters was packed wall to wall with several hundred newspeople.

Dr. T. Keith Glennan, NASA administrator, got up on the stage, where the seven of us—wearing civilian suits—were seated in alphabetical order behind a long table. Each of us had a name plate and microphone in front of him.

"Ladies and gentlemen, today we are introducing to you and to the world these seven men who have been selected to begin training for orbital space flight. These men are here after a long and unprecedented series of evaluations which told our medical consultants and scientists of their superb adaptability to their upcoming flight.

"It is my pleasure," Glennan went on, "to introduce to you—and I consider it a very real honor—the nation's Mercury astronauts!

"Malcolm S. Carpenter.

"L. Gordon Cooper, Jr.

"John H. Glenn, Jr.

"Virgil I. Grissom.

"Walter M. Schirra, Jr.

"Alan B. Shepard, Jr.

"Donald K. Slayton."

Applause erupted throughout the room as reporters and photographers stood, putting down their notebooks and cameras in order to clap vigorously with both hands. The applause continued for a surprising length of time. Some of us grinned sheepishly at each other.

It's as if we have already done something, I thought. All we had done was have our names on a list and take some tests. We hadn't even flown a mission yet.

The first question directed at us by a reporter threw most of us for a loop. It wasn't about our military or flying backgrounds; he wanted to hear from each of us what our wives and children had to say about our being selected as astronauts.

I don't remember what I said. Whatever I mouthed, it came from behind the mask of a career officer; a few platitudes, then shutting up and hoping like hell that no one knew the truth about my marriage.

After I finished, it was John Glenn's turn. He jumped on the question like a hungry mountain lion that had just spotted dinner: "I don't think any of us could really go on with something like this if we didn't have pretty good backing at home, really. My wife's attitude toward this has been the same as it has been all along through my flying. If it is what I want to do, she is behind it, and the kids are too, a hundred percent. . . ."

I looked at Glenn and smiled, thinking, *Who* is *this Boy Scout?*

He was a freckle-faced Tom Sawyer type with a sunny smile who, I quickly came to realize, was awfully good at charming his way in or out of about any situation.

I was mighty glad that I came before and not after Glenn, like my old buddy Gus Grissom, with whom I had attended some air force schools and done some hunting and flying in our spare time. Gus was not one to waste words, and each time he had to field the same question right after Glenn had scored about every point imaginable with it, poor Gus came off as distinctly inarticulate.

One of the next questions to the group was about religion. *When are we going to get some questions about flying and space exploration?*

It was as if John Glenn had been waiting for the religion question all along. He uncorked a spiel about God, country, his experiences teaching Sunday school and being on church boards, and a bunch of other saintly stuff that the press ate up.

And when he was finished, it was Gus's turn.

I turned to look at Gus, who seemed to be suffering from indigestion.

"I am not real active in church, as Mr. Glenn is."

Poor Gus.

So it went for two hours as we sat under the glaring TV spotlights and answered questions from the news media. I was startled and made uncomfortable by all the attention, as were most of the guys. Up to then we had been anonymous military pilots at midcareer. It was a taste of what the next several years would be like. We woke up the next morning on the front pages of newspapers, and shortly thereafter on the covers of national magazines. From then on we were never anonymous again.

After that, we were thankful to get to work at NASA—astronauts, technicians, scientists, and administrators, all on a first-name basis—to get America into the space race with the Soviets, who had a big head start. There wasn't a man among us who didn't feel we could overcome that lead, but it would take a lot of work and some luck along the way.

So much of what had to be done before the first manned mission could be launched had never been attempted before. NASA engineers were busy coming up with new guidance systems for rockets and new ways to monitor temperature, pressure, oxygen, and other vital conditions aboard the spacecraft—the technology was nothing short of pioneering.

We seven traveled the country—together at times, often separately—to visit defense contractors working on different aspects of the program. At the plant that was building our electronics for the rocket booster, we showed workers how to build a clean room, free of dirt, debris, and solder balls. In fact, some of the early Atlas rocket failures turned out to have been caused by errant solder balls—shaken loose by the tremendous vibrations during liftoff—that shorted out the guidance package, causing the rocket to veer off course. Soldering techniques had to be changed for all space-related work.

We came up with a Mercury stamp—the Greek symbol for Mercury with the number 7 inside it. Any part destined for the manned program, no matter how small, received the stamp—meaning that everyone from the designers to the assembly line workers realized that a man's life depended on the quality of their work. Our message was the same wherever we went: "Do good work. A man's life is riding on you."

The astronaut office at Langley Air Force Base in Virginia, where NASA was headquartered, consisted of one big office with eight desks: one for each astronaut and one for our crackerjack secretary, Nancy Lowe, who was only seventeen years old and just starting her twenty-five-year NASA career. Nancy was extremely well organized and could type more than a hundred words a minute. She handled all our letters and reports, and it was a real workload, with seven bosses, each of whom considered himself a leader—seven alpha males in the same pack.

Lots of time was spent kicking things around in that office, discussing and evaluating, quietly agreeing and loudly disagreeing, until we came up with a consensus. When a decision came out of the astronaut office, where there were only seven votes possible, it was always a single decision. There were no minority and majority decisions.

One of our most lively debates had to do with whether or not the spacecraft should have rudder pedals like an airplane to control its yaw, or side-to-side motion. In an aircraft, the yaw was controlled with rudder pedals and the pitch and roll were on the control stick. Deke Slayton and I felt that we should keep the control system we were most familiar with and have rudder pedals installed. Why make us fly the spacecraft differently than we had flown planes for years? But four of the guys felt very strongly that the rudder pedals would take up room at our feet and add to the complexity of flight-system installation. They wanted to put all three axis controls on the flight stick. Wally Schirra could see benefits to both, and rode the fence on the issue. The engineers just shrugged—they could go either way, and left the final decision up to those of us who would be doing the flying. We all did a lot of soul-searching, and Deke and I were finally convinced to go with the three-axis control stick. It was the right decision.

We weren't just mouthpieces or pilots milling around a hangar waiting to fly. We were involved in all aspects of the program, and there was a job for everybody.

Scott Carpenter took on communications and navigational aids, something he'd done in the navy. An intense and sensitive guy, Scott had a bit of poet in him and spent a lot of time thinking about and discussing the larger ramifications of space flight and what it might mean for mankind. In terms of his flying background, most of his time was in large propeller aircraft, with limited jet hours, which in this group of hotshot fighter jocks made him the odd man out at times. He was a great swimmer and diver, and the best dancer of the group—he did a mean Twist.

Al Shepard had experience with ships and knew a lot of navy brass from his naval career, so he concentrated on the worldwide tracking range and assembling the recovery teams we would need to pull us out of the water after a flight. In some ways Al was the most complex of the original astronauts. He seemed to have two distinct personalities, one the charming and beguiling jokester who introduced José Jimenez—comedian Bill Dana's popular alter ego—and his "Please don't send me" astronaut act into our everyday lives; the other that came out when the chips were down and was so competitive as to be ruthless. We all knew to watch our backs when *that* Al was around.

John Glenn of the marines had been involved in aircraft design and was a natural to work on cockpit layout and design of the instrument panel in the spacecraft. The oldest of the astronauts at thirty-seven, John had set a transcontinental speed record two years earlier in an F–8 Crusader. John would turn out to be a great and loyal friend, even if he was our "clean marine." We all knew he was destined for politics; we figured that one day he would be president. He was the only Mercury astronaut not to move his family to Langley Air Force Base. He kept his wife and two children in Washington, D.C. Whenever he went home, he made a point to visit all the NASA bigwigs and various congressional officials. As I said, John was a born politician.

NASA decided we would use a highly modified version of the U.S. Navy pressure suit for our space flights, so Wally Schirra, a navy man and Annapolis graduate, started work on adapting the life-support system that would be needed to keep the pilot alive and comfortable. Wally was outgoing and congenial and loved to make people laugh, but he was formidable when the time came and could focus with engineer-like tenacity on the technical intricacies of space flight.

Gus Grissom, who had experience in technical engineering, was assigned to help develop the automatic and manual control systems we would use to fly the spacecraft. Gus was my closest friend in the astronaut corps. We had been classmates at the Air Force Institute of Technology in Ohio and test pilot school at Edwards, and he was a heck of a nice guy, with whom I enjoyed working and playing. He was a little bear of a man and a country boy at heart, but when it came to flying he was steady and no-nonsense. Of all the astronauts, Gus would have been my first choice to fly my wing.

Deke Slayton, the other air force "blue suit" in the mix, was an experienced engineering test pilot who knew the technical details of his job thoroughly. He pitched in to study the Atlas rocket, which we would use for the first orbital missions. I had flown with Deke at Edwards and knew him to be a capable pilot. When it came to the space program, his ability to analyze a complex problem and articulate his ideas about it made him an invaluable member of the team.

As for me, I considered myself a pretty good stick-and-rudder man. Controlling an aircraft was what I did best in life. Flying came as naturally as breathing and eating to me, and if I sometimes acted as if I didn't think anyone could outdo me in the air, well, that's how I felt. Modesty is not the best trait for a fighter pilot. The meek do not inherit the sky.

Since I had a lot of propulsion work in my background, I headed out to the Army Redstone Arsenal in Huntsville, Alabama, to learn more about the rocket boosters we would use for the two suborbital flights. My work there ran the gamut: configuration, trajectory, aerodynamics, countdown and flight procedures. The Redstone was really an improved V–2, the rocket the Germans had used for blitzing London during the war. The designer of the

V–2, Dr. Wernher von Braun, who came to the United States after the war, was now a key player in our space program. Making the boosters compatible with our spacecraft would take coordination and communication between several civilian contractors, the military, and NASA. As an engineer, I could talk the language of other engineers. And since I was hoping to *ride* the finished product, I really got immersed in the problems and had a keen interest in seeing that everything went smoothly.

So we all had our jobs to do, and the work was parceled out evenly among the seven of us. For the next twenty-one months, we each went our own way, trying to learn everything about our specialty, then reporting back to the others at regular intervals.

We also spent a lot of time training in the flight simulator and centrifuge. They once took me up to 18 Gs in the centrifuge—meaning that I weighed eighteen times my normal 150 pounds, or approximately 2,700 pounds. It felt as if a Mack truck had been parked on my chest, making it extremely hard to breathe. The biomedical people were testing the limits of human tolerance, trying to figure out at what point an astronaut would go into unconsciousness, caused by gravity preventing blood getting to the brain. This happened much more quickly, we found, when the pilot was sitting upright, when there was further vertical distance between the heart and the brain—information that went into the design of our flight couches, which allowed the astronaut to be lying down during times of heavy G loads, such as launch and reentry. At 18 Gs I had still been able to answer a light signal and push buttons, but just barely. I lost my vision almost completely, and I stepped away wet with blood from broken capillaries up and down my arms, legs, and back. The only thing that could be done for me was to douse the cuts with alcohol to ward off infection. I was sore and black and blue for a couple of weeks. After that they decided not to exceed 15 Gs in training. That was plenty, given that 8 to 10 Gs were thought to be the maximum load under normal flight conditions. Eventually it was decided that 11 to 14 Gs was the maximum load at which a man could handle an intricate control problem.

One part of our training that was being overlooked was flying, and none of us were happy about it. We were doing a lot of commercial flying, going from city to city on our various NASA assignments, waiting at air terminals and claiming our luggage like any other passengers. We were fighter pilots and we longed to get back to flying. We saw a dual purpose: we'd get where we needed to be quicker while keeping up our proficiency. But NASA had no jets of its own, and although we had repeatedly asked that something be done about this oversight, our request hadn't made it to the top of any administrator's priority list.

One day, New York Congressman Jim Fulton, known for ferreting things out, showed up at Langley. I went to lunch with him. He wanted to know all about how our training was going, then asked, "And what about flying?"

I told him the truth. "We're not flying. We don't have any planes."

It was so bad, I said, that in order for us to keep our flight pay—an extra $145 a month, which required at least four hours of in-command flight time per month—we had been going out to Langley and getting in line behind deskbound colonels and generals who were trying to keep their flight pay too.

The congressman seemed shocked. "But I thought flying was a big part of an astronaut's training."

"It is," I said. "Or *should* be."

The next day, the House Committee on Science and Astronautics began an investigation into why the astronauts weren't flying. The higher-ups at NASA were none too happy, but my fellow astronauts were delighted. Within weeks we had our own F–102s—on loan to NASA from the air force. We had some single-seaters and a two-seater. The F-102 was great for us because, while it was a hot plane, it could get in and out of the shorter civilian airfields that we needed to go to and required only light maintenance.

In any research and development program where the state-of-the-art is being pressed forward, there are always unknowns—the what-if factor. Project Mercury, with its purpose to investigate man's capabilities in space, was such a program.

Extreme efforts were made to ensure operating reliability in the interest of mission success and pilot safety. But we were dealing with mechanical, manmade pieces of equipment, which even in the most ideal situations were subject to malfunction. For this reason, many redundant or backup systems were built into the Mercury spacecraft and related equipment. From prelaunch until safe recovery, hundreds of different possible contingencies were anticipated. A few examples:

IF an impending catastrophic failure was indicated between liftoff and escape tower separation (which pulled the spacecraft away from the booster about 140 seconds after launch), the automatic abort-sensing and implementation system (ASIS) could automatically initiate an escape sequence to remove the spacecraft from the booster, or the escape system could be fired by ground command from the blockhouse, or by ground command from the Mercury Control Center, or by the astronaut in the cockpit.

IF the spacecraft did not automatically separate from the launch vehicle at booster burnout, separation could be initiated by ground command or by the astronaut, who could initiate separation manually from the cockpit.

IF the spacecraft pressure vessel developed a leak during flight, the astronaut's full pressure suit would automatically inflate to five pounds per square inch to provide a second closed environment.

IF the system that provided oxygen to the astronaut's suit failed, an emergency supply would automatically cut into the circuit, or the astronaut could start the alternate system.

IF the astronaut's primary ultrahigh-frequency voice link with the ground tracking network failed, he could switch to a second UHF or a high-frequency channel.

IF the astronaut's microphone failed, a second mike in parallel with the first would automatically begin to operate.

IF all voice link systems failed, the astronaut could resort to a Morse code key in the cockpit and use the telemetry transmitters and frequencies to send messages back to the tracking network.

After designing and installing these backup systems and hundreds more like them, we believed, finally, that we were ready to go.

On January 19, 1961, the day before the inauguration of President John F. Kennedy, who would have so much to do with getting Americans into space and to the Moon, NASA director Robert Gilruth met with the seven of us in the astronaut office. What he was about to tell us, he cautioned, had to be kept in the strictest confidence.

Previously we had been told that all seven of us would undergo the same exact training, and that on the eve of the first manned flight, the name of the astronaut who would take the flight would be announced to us and the world. Gilruth now said he had thought better of that—it stood to reason that the prime pilot should have preference on the flight simulators and in other aspects of training. At the same time, he didn't want to put unnecessary public pressure on the prime pilot. So it had been decided to go ahead and select the prime pilot for the first mission, scheduled to be flown in four months, and also two backup pilots. The press and public would be given the three names and told only that the prime pilot would be one of these three men.

"This has been a difficult decision," Gilruth said. "All seven of you have worked long and hard. Any one of you would make a capable pilot for the first flight."

We all wanted him to just spit it out. All of us would have

liked to be the first American in space. We knew that selecting the pilot for the first mission was going to be largely a committee decision. Although we weren't sure who was on the committee, we knew that Gilruth was one of the influential voices that would be heard. We also weren't sure of the criteria for the decision. What would it be *based* on?

In the past twenty-one months, we had done so little flying that we had certainly not been able to show anybody what we could really do in the air. So how would they make their choice? Based on how we did on the centrifuge, or how we might be expected to handle the press and public afterward?

"But it's become necessary to make this decision," Gilruth went on. "And it's been decided that the prime pilot for the first flight will be—Alan. His backups are John and Gus."

Al looked up, fighting the obvious urge to break out into a big grin.

One by one, we went over to him and shook his hand.

Al Shepard would be the first American in space.

For nearly two years we had worked together like brothers—fiercely competitive, yes, but at the same time very supportive and wanting to see the other guy succeed in every way. That would not change now. Besides, I reminded myself, there would be plenty of other good missions down the road—longer and even *better* missions.

The first U.S. manned suborbital mission was delayed for two months after the January 31, 1961, flight of a thirty-seven-pound chimpanzee, Ham, went badly.

Although Ham survived and was to live to a ripe old age, he endured 18 Gs during reentry and, due to a faulty electrical relay, landed more than a hundred miles off target. By the time the recovery crew reached him, the capsule was on its side filling with water. The nearly drowned chimp emerged in a nasty mood, and I couldn't blame him.

Al Shepard was convinced that he could have made any necessary corrections and had a better flight than Ham. But to Al's dismay, it was decided that another test of Redstone with an unmanned Mercury spacecraft was needed before proceeding with the first manned shot. Two weeks later, that final test flight went well.

Had Al flown his mission on the originally scheduled date of March 12, 1961, he would have beaten Yuri Gargarin into space by one month. Instead, we found ourselves further behind the Soviets before we even got started, and Al became the *second* man in space—twenty days after Gargarin aboard his *Vostok 1* spacecraft. It was something that still irritated Al years later. "We *had* 'em by the short hairs," I heard him say a number of times, "and we gave it away."

The first attempt to launch Al was scrubbed two hours before launch due to cloud cover. Three days later, on May 5, 1961, I was serving as CapCom in the blockhouse near the launch pad for the third attempt—following still another weather scrub the previous day. After four hours of frustrating countdown holds, Al was still strapped into his spacecraft on the pad. Everybody was antsy to get going, but one small glitch after another kept popping up.

There had been a failed attempt, eleven days earlier, to launch with an Atlas booster a Mercury spacecraft carrying a "mechanical astronaut"—a bunch of black boxes in place of a pilot. The Atlas, which failed to pitch over into the proper trajectory, was destroyed by the range safety officer forty seconds after liftoff. However, the Mercury capsule's escape rockets worked flawlessly, separating it from the errant rocket, and the spacecraft landed underneath a billowing parachute in the Atlantic just two thousand yards north of the pad.

We all knew the stakes were much higher this day.

The delay took its toll not only emotionally but in another way.

"Gordo!" Al said urgently during still another countdown hold.

"Go, Al."

"Man, I gotta pee."

"You *what*?"

Al had been sitting atop the slim black-and-white Redstone rocket for so long that the need to urinate—after his morning coffee and juice at breakfast—was becoming overwhelming. As the suborbital flight was scheduled to last only fifteen minutes, no one had thought it necessary to equip Al or the Mercury spacecraft with a urine collection system.

"You heard me," he said. "I've got to pee! I've been in here forever. The gantry is still right here, so why don't you guys let me out of here for a quick stretch?"

I turned around and told the experts the latest glitch.

"No way, Al," I reported back. "They say we don't have the time to get you out of there and start all over. You're in there to stay."

"Gordo, I've got to *go*!" Al shouted back. "I could be in here a couple more hours. By then, my bladder's gonna burst!"

"They say no."

Al's famous temper was soaring. "Well shit, Gordo, we've got to do *something*! Dammit, tell 'em I'm going to let it go in my suit!"

Nobody thought that was a real good idea.

"No good, you can't do that," I told Al. "The medics say you'll short-circuit their medical leads."

"Tell 'em to switch off the damn power," Al snapped.

I turned to the experts. There was more discussion—the only wires that urine was likely to come in contact with were low-voltage leads to the medical monitors attached to Al's body. There was even a sponge mechanism inside the suit designed to sop up any extra moisture, such as perspiration. The experts finally gave the go-ahead.

"OK," I said. "Go to it, Al."

Given Al's position in the spacecraft, the urine pooled in the back of his space suit without soaking any of the sensors on and around his chest area. Later, in his best José Jimenez accent, Al would proclaim himself "de first wetback in space."

After that unscheduled pee, the first U.S. manned launch was textbook all the way. Al's *Freedom 7*—the number stood for factory model number 7, the spacecraft delivered by the manufacturer for Al's flight, not for the seven original astronauts, as most people thought—reached a maximum speed of 5,180 miles per hour, soared to an altitude of 116 miles, and landed 302 miles downrange. The flight lasted fifteen minutes and twenty-two seconds; Al experienced a five-minute period of weightlessness and reported no ill effects.

Two weeks later our youthful president, John F. Kennedy, told the nation that we would "put a man on the Moon and return him to Earth safely before this decade is out." It is still the best mission statement of all time, but back then there were a lot of gasps behind closed doors at NASA. Here we were with a total of only fifteen minutes of U.S. manned space flight experience, and we'd never put a man in orbit. President Kennedy's deadline was two years ahead of the most realistic timetable. We didn't have the huge rockets that would be needed to get to the Moon or the spacecraft in which to fly there. We didn't even know how to navigate our way there and back. Don't get me wrong, I knew it could be done if we put our minds to it—the question was whether all the hardware we needed could be built and tested in time. In spite of the gasps at NASA, everyone was *elated* to get this kind of public backing from the president on such a big program. Nonetheless, a London bookmaker promptly placed thousand-to–one odds against the Moon landing on JFK's schedule. (We should all have taken the bet.)

Gus Grissom was selected to fly the second suborbital mission—a rerun of Al's fifteen-minute flight—and went up aboard

Liberty Bell 7 on July 21, 1961. (Gus kept the number 7 to "honor the team," and thereafter every Mercury capsule had that symbolic number in its name.) Although he flew a flawless mission, Gus ended up being slaughtered by the media when his capsule, loaded with valuable scientific data, sank in the ocean during recovery efforts. It was the first time in his flying career that Gus had ever lost an aircraft—and it was the only unrecovered U.S. spacecraft—although he and I had nearly bought the farm together earlier in the program.

After the first test of an Atlas rocket carrying an unmanned Mercury capsule went up in flames forty-five seconds into the launch, it was decided to watch the next launch closely from aloft to see if we could spot any obvious problems. Gus and I volunteered to trail the launch in two delta-wing F–106s, keeping the big Atlas company as it gathered speed after liftoff. We would have real grandstand seats. Gus was to approach at five thousand feet, ignite his afterburner, and climb up in a spiral alongside to observe this early phase of its flight. I would take over from the fifteen-thousand-foot level and continue observing the big bird. We both had cameras around our necks with which to take pictures of the rocket at different stages in the flight. I was doing 1.3 Mach and coming around on the Atlas, expecting it to pitch over and begin its arching flight out over the Atlantic. But instead, it kept going straight up. At that instant, the range safety officer in Mercury Control pushed the abort button, which was a proper call on his part. The only problem: no one remembered to warn us. The next thing I knew, the escape tower fired, pulling the capsule right past me—at least *that part* worked. The tower and capsule missed my aircraft by no more than fifteen feet, and a fireball was heading up after me. Reacting automatically, I pulled up and over and pointed the nose down, leaving the afterburner on, and ran like a son-of-a-gun to get out of the debris that was falling in chunks. Gus came through the flames and debris below me.

Somehow, neither of our planes picked up so much as a single metal fragment. It was the last time any of us tried to follow a rocket in a plane.

From my experience flying with him, I knew Gus was a great pilot. When he returned from his space mission and insisted that he hadn't blown the hatch early—before navy frogmen had a chance to right the spacecraft and put a floatation collar around it—I believed him. I knew that if he *had* screwed up, Gus would have been the first to admit it. Instead, he assured us, "I was just lying there minding my own business when—pow!—I saw blue sky and water coming in. It was the biggest shock seeing that hatch go off."

He was lucky to get out at all with three thousand pounds of sea water pouring through the open hatch. Once he did, he found himself flailing around in the ocean under the roar of helicopter blades as the recovery team concentrated on trying to snag the sinking spacecraft before turning their attention to him. "I'm drowning, you bastards!" he yelled up at the helicopter crew, while they waved back and smiled.

That Gus nearly drowned (his helmet was recovered floating in the ocean next to a ten-foot shark) didn't seem to matter as much as the rampant speculation among the news media over whether he had blown the hatch early. He wasn't treated to a parade after his flight, or official celebration from the White House on down, as Al Shepard had been. Although this would undoubtedly have been the case for anyone flying the somewhat anticlimactic second suborbital flight—the media were already getting excited about the upcoming first orbital mission—Gus and his wife, Betty, couldn't help but feel snubbed by the press, public, and politicians.

Two weeks after Gus's mission, NASA learned conclusively that Gus had not screwed up when they discovered the real culprit: a design problem with the ignition pin for the hatch, an explanation that was buried on a back page of the *New York Times*.

Although Gus was subsequently cleared by an official accident review panel, he would continue to be haunted by the incident, thanks to the media. The 1983 movie *The Right Stuff* portrayed him as a panicky oaf who "screwed the pooch" by losing his spacecraft. Most importantly, though, NASA never held the incident against him. In fact, they later gave him the first two-man Gemini flight and scheduled him for the first three-man Apollo mission. They would have done neither if they believed that Gus had panicked in any way. In July 1999, *Liberty Bell 7* was recovered from the ocean floor—three miles down, deeper than the *Titanic*—in a salvage project underwritten by a television network. I don't know what they will eventually find in the way of physical evidence aboard the spacecraft, but I know that when John Glenn and Wally Schirra blew their hatches on their Mercury missions—signaled to do so by navy frogmen after they'd placed the floatation collars—they ended up with cuts on the palms of their hands from the handle of the hatch-release mechanism snapping back and getting them good. Gus had no such wounds.

The third Mercury flight and first U.S. orbital mission, flown by John Glenn on February 20, 1962, aboard *Friendship 7*, put us in the space race for keeps.

As with the first suborbital flight, this mission was preceded by the launch of a chimpanzee, Enos. Scheduled for three orbits, the spacecraft was brought back after only two orbits due to the failure of an attitude-control thruster jet and the overheating of an inverter in the electrical system. Making matters worse for Enos, each time he responded to a light during the flight and pushed a button as he'd been trained to do, he received not a banana pellet, as he should have, but an electric shock. He was a very pissed-off chimp by the time the hatch was opened. Engineers decided that both of the major problems could have been corrected had an astronaut been aboard, and John was confident that, with or without banana pellets, he could have flown the mission.

Some sixty million people viewed John's launch on live television. His spacecraft made its three-orbit flight in four hours and fifty-five minutes. Prior to the flight there was concern about the physiological effects of prolonged weightlessness. John reported no debilitating or harmful effects, and in fact found the zero gravity conditions very handy in performing his tasks.

A faulty switch in the heat shield circuit indicated that the clamp holding the shield, designed to keep the spacecraft from burning up during reentry, had been prematurely released—information later found to be false. During reentry, the retropack (containing the spent retro rockets used to maneuver in space) was not jettisoned, as the flight plan called for, but retained as an added measure to help hold the heat shield in place in case it had somehow been loosened. That caused some fireworks as the spacecraft reentered the atmosphere and the retropack disintegrated in the intense heat.

John's spacecraft landed in the Atlantic eight hundred miles southeast of Bermuda and was recovered by a navy ship after being in the water about twenty minutes. With the success of the flight, we were beginning to achieve the basic objectives of Project Mercury: putting a man into Earth orbit, observing his reactions to the space environment, and returning him safely to Earth in a place where he could be readily found.

After his flight, John, perfectly at ease as always when it came to dealing with the crush of media attention, became a wonderful spokesman not only for NASA but for America. He did a tremendous amount of good for the program.

The next flight was to have been Deke Slayton's, but he was abruptly grounded for having a detectable heart murmur at high Gs in the centrifuge. We thought this was extremely unfair, because anyone could be made to have a heart murmur at high enough Gs. Deke was ready to quit, and the rest of us were so upset that we came close to resigning en masse, which would have left America's space program with no astronauts. We finally talked Deke into staying on as head of the astronaut office, a new post in which he would represent our mutual interests. It wasn't a charity appointment. At the time, there was serious talk of bringing in a general or admiral to fill the new post, but we all agreed that we didn't need some outside weenie coming in and telling us what to do.

Deke's backup, navy pilot Scott Carpenter, was given the next orbital mission, although he didn't have much time to practice for it in the simulator as prime pilot. Scott was launched on May 24, 1962, aboard his *Aurora 7* spacecraft and replicated John Glenn's three-orbit mission. Things didn't go well almost from the beginning. Scott used more fuel for the attitude-control thrusters than planned due to a faulty automatic stabilization system, and had to drift for most of his final orbit. Scott got caught up in some scientific experiments and wasn't ready for his retro-fire. He fired late, and overshot the recovery area by two hundred and fifty miles. After spending more than an hour inside the hot cabin, he wiggled out of the narrow top of the tower rather than blow off the hatch and risk the capsule's sinking. He spent more hours on his raft before being picked up by a helicopter. NOT CONFUSED: CARPENTER blasted one headline. Scott, who knew he had screwed up, took some heat. He never flew in space again, and when the opportunity arose to become involved in the navy's Sea Lab research project, he quietly resigned from NASA and ended up doing a fantastic job in the undersea exploration program.

Next up was Wally Schirra, another top navy pilot. Wally believed in training himself hard, and was very precise. After his launch on October 3, 1962, aboard *Sigma 7*, Wally experienced problems with his pressure suit that almost brought him back to Earth after only one orbit, but he was able to solve the difficulties in flight and sailed on for six orbits. His mission was considered a textbook engineering flight and went so well that NASA decided to advance the flight schedule.

The original Mercury launch schedule had provided for two six-orbit missions before trying for the first one-day mission, but after Wally's flight, NASA decided to jump to the one-day mission—and it was to be *mine*. Eighteen orbits were the goal, but if the flight met all expectations, Mercury Control would extend the mission to twenty-two orbits—leaving me in space for approximately a day and a half.

New state-of-the-art design modifications had been made to my spacecraft to accommodate life-support and functional demands for more than a day in space. Since weight was extremely critical, attention was first given to removal of any unnecessary hardware and systems. The periscope (nothing more than a sightseeing device) had already been removed from the spacecraft for Wally's flight. For my flight they undertook a complete weight reduction program—measuring the payload not in pounds but in tenths of pounds. The manual rate stabilization and control system—like an autopilot that kept constantly adjusting the spacecraft's attitude with thruster rockets—turned out to be unnecessary and a glutton for fuel, so it was removed. A lighter-weight pilot's couch was installed. Nine extra pounds of water, about a gallon and a half, were added to the cooling system, and four pounds of additional liquid oxygen augmented the breathing stores. I was scheduled to drink voluminous amounts of water for urine analysis, which NASA physiologists thought would be the best way to monitor chemical reactions that might occur within the body during a period of prolonged weightlessness, so five additional pounds of drinking water were added. Electrical battery power was doubled. A television camera—the first to go into space—was installed to prove the feasibility of TV in space and to record ground scenes as well as onboard life.

I had been given three main priorities for the flight: prove that a pilot could go off automatic control and fly the spacecraft efficiently; capably manage the use of all onboard consumables, such as fuel for the attitude-control thrusters, electricity, oxygen, and water; and conduct scientific experiments to aid development of rendezvous techniques, which would be vital in the upcoming lunar missions.

NASA had another top priority for the mission: "determining the effects of extended space flight on man." Many doctors and scientists were convinced that the pooling of blood in a person's extremities during zero gravity would preclude space flights of more than a few hours. The sickness that the Russian cosmonauts were reported to have experienced on their long-duration flights increased the concern.

When it came to naming my spacecraft, I felt a responsibility to find the right name. I knew that an awful lot of thought and symbolism had gone into the earlier names, and I finally selected *Faith 7*, which was painted on the side of the spacecraft. The name would symbolize my faith in the launch team, my faith in all the hardware that had been so carefully tested, my faith in myself, and my faith in God.

Four years and five months after volunteering for the manned space program in that crowded NASA briefing room, I was about to get my chance. But not before a close call—nearly getting bumped from the mission at the last minute by an angry Walt Williams, Project Mercury operations director and NASA's number three guy.

I had been the one to get angry first. When I arrived at Hangar S the day before launch, I saw to my disbelief that a last-minute modification had been made to my custom-made pressure suit. The medical people had cut a hole in it and inserted a metal fitting in the chest for an additional monitor, which they could automatically inflate any time they wanted to during the flight to check my blood pressure. This violated a firm rule we had against doing *anything* to an astronaut's pressure suit just before a flight. What if the new fitting leaked? I might not discover that my suit wasn't holding pressure until I was in space and needed it.

I hit the roof, and was still venting when I took an F–102 aloft for what had become a day-before-launch ritual to allow an astronaut—feeling overworked and overprepared about now—to have a little fun. I pushed the supersonic fighter hard through loops and dives. Then I dove for the deck. As I zoomed past Hangar S only a few feet off the ground, Williams happened to step from a doorway, only to see himself eyeball-to-eyeball with a streaking jet fighter. At that instant, I lit the afterburner. I heard later that Williams dropped the stack of papers he was holding and grabbed at his throat as if to keep his heart from leaping out of his chest. He went from quivering to an uncontrolled rage. The condemnations, which came to me secondhand because Williams was unwilling to confront me directly, went on until well into the evening. A stern man by nature, Williams insisted in no uncertain terms that I was *off* the mission and ordered Al Shepard, my backup, to take the flight. Al had been aggressively seeking a chance to get into space since his fifteen-minute suborbital flight, which now paled in comparison with the longer missions being flown, and was very happy to do so. This came as no surprise to me because I knew this side of Al, who was not to be trusted when it came to anything that he considered in his best interests.

Williams had long believed I was not taking things seriously enough. He'd come from being chief of operations at Edwards for NASA's rocket plane programs—X–1, X–2, and X–3—and was accustomed to dealing with serious and studious research types, not fighter jocks. I guess I just rubbed him the wrong way, because much earlier he'd raised with Deke Slayton the subject of giving Shepard the nod over me. "Walt thinks you're a hot dog and maverick and that you bend the rules too much," Deke had told me. I'd given Deke my best "Who, me?" look. Had it been all those speeding tickets in Cocoa Beach? Or my racing at Daytona International Speedway? Or maybe running the world-class speedboats? Sometime later Deke told me that President Kennedy, whom I'd met more than a dozen times, had gotten wind that I might be passed over and had made it clear to NASA that he would have none of it. It's good to have friends in high places.

Now I'd gone and given Williams the opening he was looking for.

The other five astronauts backed me, saying that the medical people had caused the problem in the first place by tampering with my suit at the last moment. But after four years of nonstop training and preparations, I found myself facing the possibility that I might lose my shot. The wait was agonizing.

Then, about ten o'clock the night before the launch, I got the word that Walt Williams had relented.

The mission was still mine.

LIFTOFF!

By launch time the thin-skinned Atlas was little more than a venting bag of combustible gases. Its thin stainless steel wall required three pounds per square inch of pressure just to keep its shape, even during transfer on a flatbed truck. Without it the Atlas would literally fold up.

The booster I was sitting atop was designed to explode under control, lift its payload off the ground, and kick it into orbit one hundred and fifty miles up.

It promised to be one hell of a ride.

The Sun had come up, and the morning was cloudless. The countdown progressed without incident until T-minus eleven minutes and thirty seconds, when a technician inside Mercury Control spotted a possible problem with the guidance equipment.

A hold was called, and I waited.

I was lying on my back on the contour couch, with my feet jackknifed above me—as if I had sat in a chair and then someone had placed the chair on its back. Mercury was so cramped you didn't climb in so much as you put it on. Built by McDonnell

Aircraft in St. Louis with the help of four thousand suppliers, including 596 direct subcontracts from twenty-five states, the spacecraft contained 750,000 parts and seven miles of wiring.

Twenty-four inches away from my face was the instrument panel, with more than a hundred gauges, dials, meters, and switches. The left-hand side of the cockpit held instrumentation for flight events, navigation, and spacecraft control, while the right side contained instruments for the environmental and communications systems.

I was wearing a full pressure suit, complete with helmet and gloves. Although in later manned flights we would have more confidence in the cabin pressurization system and remove our suits during space flight, in Mercury we left the cumbersome suits on all the time in case there was an emergency depressurization, at which point the suit would automatically pressurize, providing a livable environment.

The suits were not real comfortable, in spite of multiple fittings by experienced tailors who laboriously custom-made them for each astronaut. They fit like a glove, and that was part of the problem. They had to fit tightly or during pressurization there would be a lot of excess air inside the suit, making movement more difficult, but the tight fit, especially when you had the suit on for a lengthy time, tended to cause chafing and sore spots on the arms, legs, and torso.

The potential glitch never materialized, and the countdown resumed.

At T-minus nineteen seconds, a momentary hold was called by Mercury Control to determine whether all systems had gone into automatic sequencing, as planned. When everyone was convinced that we were a "go," the hold was lifted—

—and the candle was lit.

Over my headset came Wally's excited announcement: "We have liftoff!"

As I had done hundreds of times on the centrifuge—outfitted with a complete Mercury control panel—and in the flight simulator, I reached up and started the onboard clock, which would keep me on schedule with all my experiments, chores, and flight operations during the mission. As a backup, I also started the stopwatch function of my Omega Chronograph wristwatch.

"Roger, liftoff," I said. "The clock is started."

My immediate sensation was that someone had pressed the pedal to the metal of a hot car, only it was unlike any powerful machine I'd ever ridden in—including dragsters, Indy cars, and high-performance racing boats. This one just kept *on going*.

Accelerating . . .

And, unbelievably, *accelerating more*.

It had no equal in my frame of reference, and I had flown the hottest fighters in the U.S. arsenal. The weight of the rocket on the launch pad was 260,000 pounds—fully 200,000 pounds of that in rocket fuel alone. Now, with its fuse lit, the Atlas was consuming fuel at a rate of 2,000 pounds per second.

"Feels good, buddy," I told Wally. "All systems go."

As I kept a close watch on the instruments and recited readings to Mercury Control, I felt the steady buildup of G-forces.

It took about twenty seconds for the rocket to surpass the speed of sound (seven hundred miles per hour), known as Mach 1. The vibrations, which had started at liftoff and increased as the rocket continued upward, smoothed out once I reached transonic speed. Now that I was out in front of the noise wave, things quieted down considerably inside the cabin. The ride was much smoother than the centrifuge at high Gs—I didn't feel nearly as beat up.

Two minutes after liftoff, the two outboard engines cut off as planned and dropped away from the rocket. The escape tower at the top of the capsule was jettisoned with the first stage, since the spacecraft was high enough by then to go through normal parachute deployment in the event of an emergency reentry.

At the sudden reduction in thrust, I was slammed forward against my chest harness, but in less time than it took to take a breath, I was snapped back and pressed into my seat by acceleration from the sustainer engine, the third of the rocket's three main engines. We weren't at the finish line yet!

The Gs kept building, and at their peak I was straining to keep my vision.

When the spacecraft reached orbital velocity, on a path that had leveled off and was now running parallel to the Earth's surface, the sustainer engine cut off, and a series of small rockets fired to separate the spacecraft from the booster.

My spacecraft was inserted into orbit at almost the exact speed required to achieve a circular orbit of the Earth. The centrifugal force of the spacecraft's forward motion would keep it literally free-falling around the Earth on a course no closer than 100 miles (perigee) and no farther away than 165 miles (apogee).

I figured I must have set a new speed record for all mankind: *0 to 17,546 miles per hour in five minutes flat.*

When the Gs hit around 8 or 9 they stopped, the pressure against my chest and body disappeared, and I was left hanging there. I had gone "through the gate," as we called it: I was in orbit and weightless.

I recognized some sensations from all the hours of practice in the centrifuge, when they ran the G-forces up and then abruptly cut the speed. But the release from being many times my normal body weight to having no weight at all when in orbit was brand-new, and something we had been unable to practice in the centrifuge. It caused a sudden full feeling in the head, due to the cardiovascular system having to work harder to pump blood under high Gs and then suddenly being caught pumping too much blood. While this didn't affect my mental or physical functions, I did feel flushed in the face for about twenty minutes until my body shifted into a lower gear and my cardiovascular system settled down and adjusted to weightlessness.

During planning of the mission, I had fought to keep my first orbit uncluttered, not wanting to become overwhelmed with work right away. I'd be in a new environment, and I wanted to be able to take a few minutes to adjust to the situation and look around before starting on what would prove to be a busy flight plan.

At separation, the spacecraft had begun its automatic turnaround. When I looked out, I could see the shell of the Atlas rocket following me, tumbling end over end as it grew smaller. It had been transformed from the world's biggest Roman candle into a burned-out hulk that would eventually plop into the Pacific a few hundred miles downrange.

I turned off the autopilot and switched to fly-by-wire, which afforded me electronic control of the spacecraft. As called for by the flight plan, I gently swung the nose around to put the heat shield in front and established a negative 34-angular-degree heat shield-upward attitude. This attitude was necessary in case I needed to fire the retro rockets—which were inside an external package mounted to the heat shield—to slow my forward speed in the event of an emergency reentry should Mercury Control find something wrong with the orbit or my spacecraft during the next few minutes.

I was heading east, toward Africa, but because I was riding backward I was looking west—back from where I had come.

A small window—about twelve by eight inches—was a late addition to the Mercury spacecraft at the insistence of the astronauts. The only way astronauts on the first two suborbital flights had been able to see out was through the periscope, since removed, which had provided only a limited field of vision.

The window—triple-thick quartz to withstand heat—had been placed in the shell of the spacecraft just above the pilot's head. It was easy enough to look out—

Oh Lord, what a heavenly view.

It was a perfect moment, when time seemed to stop. I was seeing our planet from afar for the first time, and I wanted to live within the moment for as long as possible. I had finally made it—I was in space, and already having a ball.

I could see most of the Caribbean and many identifiable islands, including the Bahamas, a spectacular sight of bright colors and pure white coral reefs. I saw the entire state of Florida, and up the eastern seaboard as far as Virginia. It all looked just the way it did on the maps, except that there was a bluish tint to everything.

In a few minutes, with all systems checking out, I was cleared to continue with the mission. I was free to change attitude now, which I did, coming back around—with the space-craft's nose facing forward—so that I could see in front of me. Before me lay the continent of Africa, filled with rich browns and golden light.

About then I realized how warm I was getting. Air friction during launch had heated the spacecraft's exterior surface to around 1,300 degrees Fahrenheit, and some of that intense heat had radiated inside the spacecraft, putting a load on the cooling system. This had also been a problem for the other Mercury flights. A water boiler gave us cabin cooling. If you set it too high it would freeze up and give you nothing; too low and you'd get no relief. I fiddled with the settings until my suit dropped about 10 degrees and the cabin temperature about 20 degrees, to a more tolerable 92 degrees and 109 degrees, respectively.

Looking out the window again, I saw John Glenn's "fire-flies." Since his flight we had learned that these were carbon and ice particles that formed from the firing of the attitude-control thrusters. They tended to cling to the outside surface of the spacecraft. John had been astounded when he hit the wall of his spacecraft and saw all these objects ("I have no idea what these little critters are"), glowing from the light of the Sun or Moon, encircling him. Every astronaut since had enjoyed repeat performances.

Near the end of my second orbit—some three hours into the mission—I activated the flashing light experiment, launching a sphere with a battery-powered strobe designed to flash once every second. The metal ball left its cylindrical housing and settled into its own orbit, near my spacecraft. I engaged manual flight control and maneuvered around, trying to locate the light out my window, but with no luck.

I went on to other things, then finally, on the night side of the fourth orbit, after yawing the spacecraft around 180 degrees, I spotted the sphere—shiny from sunlight glinting off it. It was coming up slowly from below, not more than twenty feet away, pulsating brightly every second just as advertised. Able finally to confirm the success of the experiment to Mercury Control, I became the first person to launch a satellite while in orbit. What we were to learn from this and other such experiments would prove useful in developing docking procedures for later flights.

I was less successful, two orbits later, when I attempted to release a balloon as part of an experiment to evaluate atmospheric drag on the spacecraft. Ground control indicated that their instruments showed the balloon failing to deploy from its external housing. I tried a couple more times before we all agreed it was a waste of time.

On that same orbit I reported that I could clearly spot a town in South Africa where flashing lights were being pointed at me, part of an experiment to evaluate man's ability to observe from space a light of known intensity. It was believed that this would have application to the Gemini and Apollo programs, especially in the landing phases.

One of the most astounding and unexpected features of the flight was my ability to distinguish objects on Earth. Among my many clear sights: several Australian cities, including the large oil refineries at Perth; wisps of smoke from rural houses on the continent of Asia; Miami Beach, Florida; and the Clear Lake area near Houston. While passing over the heart of America, I was able to see the entire west coast, then swing around to the east and view the whole eastern seaboard. My elapsed time traveling from one coast to the other was just *fifteen minutes*. Such experiences brought home how small our planet is in the context of space exploration.

Beginning the eighth orbit, I shut down the nonessential control and electrical systems and settled into drifting flight. At this altitude—a hundred and fifty miles up—the surface of the spacecraft was still hitting molecules of air, which had a slowing-down effect. Without propulsion to boost its speed and make up for the drag, the spacecraft, left on its own, would eventually—in a few weeks—slow down and drop from orbit. Longer orbiting missions—like the space station of the 1990s— would need to be in orbit higher up, around three hundred miles, where the atmosphere was thinner and there was less drag.

In zero gravity there is no right-side up, so it wasn't necessary to use up vital fuel constantly adjusting the attitude of the spacecraft with respect to the Earth. I found drifting to be a very pleasant and relaxing mode of flight, and came to look forward to the Earth slowly spinning by time and again to fill my cabin window with colors and patterns like a glorious living mosaic.

The flight schedule called for me to go to sleep at the beginning of my ninth orbit, but when the time arrived I was busy taking pictures of the High Steppes of India, approaching the Himalayas. It was the first time we'd ever had a spacecraft fly in daylight over the Himalayas. I had always been fascinated with that area, so I stayed awake photographing it.

NASA's position, expressed in a memo, was that "If an astronaut desires, he may carry a camera." That revealed the prevailing lack of importance placed on photography in space. John and Scott had carried Instamatics, which I guess are better than no camera at all. But as an experienced amateur photographer—going back to my childhood days, when I was always the one with the camera at family gatherings—I had been disappointed by the lack of good pictures from space. So I had brought aboard a stripped-down lightweight Hasselblad, which I'd had modified for me by the air force camera lab at Patrick Air Force Base. I also carried a light meter to test light values.

As I kept shooting pictures, I didn't say anything and stayed quiet in the cabin, hoping the guys at Mercury Control would think I was asleep. I maneuvered the spacecraft in order to take my pictures, keeping a light touch on the pulse jets so that no noticeable amounts of fuel would be expended. Otherwise Mercury Control would detect it and ask me what I was doing and why I wasn't asleep.

After another orbit of photo-taking, I decided to turn in. I had managed to get the temperature of my suit down further and was feeling very comfortable. As I passed over Asia and started back across the Pacific, I fell into a deep slumber.

I had a pretty low pulse rate naturally, and with the heart not having to work as hard in zero gravity as it does on Earth, my pulse had gone down to the forty-beats-per-minute range. Then, when I finally did go to sleep, it dropped lower.

Since no U.S. astronaut before me had been in space long enough to require a sleep period, there had never been an occasion to monitor someone asleep in space. In fact, some of the medical people weren't even sure if a prolonged sleep would be possible in space. They also weren't sure a man could sleep in space, then wake up and get his senses back in order to pick up where he left off. The first time I came whistling over the Cape, sound asleep, Mercury Control became so alarmed at my low heart rate that the worried doctors decided to have CapCom awaken me to make sure I was all right.

My biggest shock, upon being so rudely awakened, was the sight of my hands floating freely in front of me—dangerously close to the control panel.

I assured the ground that I was feeling just dandy. Before going back to sleep, I improvised by hooking my thumbs under my helmet straps to prevent my arms and hands from floating and accidentally tripping a switch. I went back to sleep for a restful six hours, proving to any and all doubters that it was possible to sleep in space. It was one of my favorite experiments.

Getting back to work on the fifteenth orbit, I checked my consumables and reported that I had used only 4 percent of my manual control fuel supply and 22 percent of the automatic control supply. Also, 75 percent of the primary supply of oxygen remained, and the reserve supply was untouched.

Mercury Control congratulated me. Al Shepard, serving as CapCom, added a personal aside: "You're sure a miser, Gordo. Stop holding your breath."

There's a reason I used less oxygen than any of my predecessors: I was the first lifetime nonsmoker in space. Some, like John Glenn, quit during the program; others, like Wally Schirra and Al Shepard, smoked right up to the launch pad. In that controlled environment it was documented that my lungs were much more efficient. My oxygen consumption was only one-third of what they had calculated it would be based on previous flights, a powerful argument for not smoking.

About then it was dinnertime: bite-sized cubes of peanut butter and chocolate, along with squeeze tubes filled with a high-calorie meat. Life-sustaining was about the best that could be said of the menu—and in some thirty-five years of commercial airline travel since, not much has changed when it comes to in-flight meals.

After the engineers on the ground decided that my spacecraft was operating "in almost unbelievable fashion," I was given the go-ahead for the full twenty-two orbits.

"Copy that," I said.

I may have sounded calm, but I felt like letting out a big "Whoopee!"

On sixteenth orbit, I swung the spacecraft around and, holding up my 35-mm still camera to the window, took pictures of the zodiacal light, a colorful glow seen on the horizon at sunset when the ions in the air are illuminated by the Sun's rays for only a minute or two. The multicolored lights were spectacular from my vantage point. The first band close to the Earth was red, the next was yellow, the next blue, then green, and finally a kind of magenta. It was a brilliant rainbow that didn't last long. After the Sun went down, the Earth turned inky dark. This series of horizon pictures was for the Massachusetts Institute of Technology. MIT planned to study the photos to determine whether the horizon could be used as a guidance reference for returning Moonships. (It was found to be one method for alignment and navigation, but later it was decided that celestial navigation—using fixed stars—was more precise.)

My textbook flight continued until the nineteenth orbit.

On that orbit, I was adjusting the cabin-light dimmer switch when a green light that measured the pull of gravity came on. It wasn't supposed to go on until reentry. I knew I wasn't reentering the Earth's atmosphere because I hadn't done anything to slow down, which I would need to do to drop from orbit.

In those days, controllers on the ground didn't have continuous communication with an orbiting spacecraft—there were only five or six places around the world where they could talk to one another. As a result, there would be blackout periods of fifteen to twenty minutes when we'd be out of touch. There were times when the radio silence was a pleasant respite from being asked ten questions a minute by ground control, but it never failed that whenever problems developed, we were unable to communicate. I was in a blackout period, *of course*, when the light came on.

Panel lights coming on for no reason were a pet peeve of mine. I'd complained so much about them, warning the technicians that one day I was going to take a hammer and knock out the offending lights, that when I climbed into the spacecraft for launch there was a little toy hammer hanging off a relay handle on the control panel.

When I was back in range, I reported the bogus light to Mercury Control.

They had already seen it on their own screens and were alarmed that I had started an unplanned departure from orbit and was already in reentry.

"Like hell," I said, a bit irritated at the suggestion.

I was instructed to power up my attitude-control system and to relay information on attitude indications present on my gyroscopes, which are spin-stabilized and provide reliable attitude references on all three axes. I did, and all telemetry implied that there had been no orbital decay and that my speed was correct.

"Stand by," I was told. "We'll run it down and see what it is."

When Mercury Control came back on the line, I was given a number of try-this-and-that directions. It was obvious, by what they were having me do, that the assembled experts in the control room were grasping at straws.

By then my electrical system was shorting out across the board, and a lot of things were going on the blink. Even my telemetry—vital information from my spacecraft that technicians at Mercury Control read on their computer screens—was dropping off their screens. They would soon not have the foggiest idea what was working or not working aboard my spacecraft.

I had no idea what was happening, and no one on the ground could come up with a cause or a solution either.

On the twentieth orbit they delivered the bad news: "The systems people have analyzed the problem. We've determined you're having a total power failure."

Total power failure.

My cooling system went next. This meant that I would be unable to regulate the temperature inside my suit or the cabin and would be unable to purify the oxygen, thereby getting rid of the carbon dioxide that would be building up. Increased temperatures inside my suit would make things decidedly uncomfortable, and eventually dangerous, but the gravest threat would come from the buildup of poisonous carbon dioxide gases. Impaired judgment would come first, then blackouts—the beginnings of brain asphyxia.

There was only one thing to do: get down as soon as possible; preferably in a planned recovery area.

My gyroscopes went out.

Then my clock stopped—the finely tuned instrument that was supposed to keep me on schedule every second of the flight.

Bit by bit, I was losing all on-board electronics.

The only reason I could still communicate with the Cape was that the radio, designed to be independent of all other systems, was hooked directly to batteries.

We all knew, without saying, that we were experiencing the worst systems failure in the more than two years that America had been sending men into space.

I was on a dying ship.

BRINGING
HER HOME

Things were beginning to stack up.

I reported to Mercury Control that I had lost all electrical power, carbon dioxide levels were above maximum limits, and cabin and suit temperatures were climbing. "Other than that," I deadpanned, "things are fine."

We were always honest about assessing what could go wrong and trying to anticipate all the possible dangers. I had long thought the worst-case scenario would be an in-flight fire that destroyed all onboard electronics and environmental systems. To prepare for such a situation, I had spent many hours in the simulator operating in the most degraded mode I could by turning off everything, then working on bringing her home.

Although I hadn't experienced a fire, the total-failure scenario I had worked on in the simulator was exactly what I now faced.

I could tell from the concern in their voices that the guys on the ground were highly concerned.

I reminded them that I had spent a lot of time on the simulator practicing this very thing. Quickly they retrieved the simulator records and saw all the work I had done. I think that gave them a little more confidence that it could be done.

We didn't have much choice. If we did an emergency reentry and I came down early, I'd lose my primary landing spot in the Pacific where the recovery fleet was waiting for me. I could end up landing in some godforsaken place where I would be stranded for a week or more while recovery teams searched for me.

With the automatic control system not functioning, I would have to reenter in the manual mode. While proving that I could go off autopilot and fly the spacecraft had been a priority of the mission, no one had ever thought it would become a matter of life or death. With my gyros out, I would have to establish my spacecraft's angle of attack using only the horizon as my reference point. Then, manipulating the hand-grip controller located next to my seat, I'd have to fire the retro rockets at precisely the right moment and hold the spacecraft steady on all three axes— pitch (vertical), roll (lateral), and yaw (side to side). When the electronic damping system was working, these corrections were automatic, offsetting for the thrusters so that the spacecraft didn't start oscillating wildly. Now, I'd have to damp out any oscillation as soon as it happened or the spacecraft would quickly spin out of control and rapidly disintegrate in the atmosphere. Although there was no time to think about the historical implications, I was attempting to become the first astronaut to fly the complete reentry trajectory under manual control.

Assuming more control over reentry was exactly what we astronauts had wanted from day one. We were pilots, after all. There was a conflicting school of thought, which held that automated and redundant systems would be more accurate and trustworthy in the long run. The issue had a history.

In the late 1950s, when America's space program was still on the drawing board, a debate had broken out about what type of people should be sent into space. High on the list were gymnasts, due to their conditioning and flexibility. President Eisenhower broke the deadlock by declaring that only military test pilots should be recruited. Even so, because of all the electronics and automation subsequently built into the Mercury spacecraft, we had sometimes been crudely compared to the chimps who had ridden atop the first U.S. rockets (the Russians used dogs). Some fellow pilots, especially those who hadn't been selected for the program—like Chuck Yeager, who hadn't met the criteria for selection, partly because he wasn't a college graduate and was too old—were left with sour grapes. They joked about us being nothing more than "Spam in the can."

This flight would put an end to all that nonsense.

My electronics were shot, and a *pilot* had the stick.

As I passed over Africa starting my next-to-last orbit, Mercury Control advised that on reentry I would be firing the retro rockets on my own count, using the second hand on my wristwatch, backed up by a countdown from John Glenn aboard the communications ship in the western Pacific.

One problem we went over was that the three retro rockets—whose nozzles were located on the outside of the spacecraft—pointed off in different directions. When fired, their thrust would place a heavy yaw on my required angle of flight if I didn't make the corrections that the electronic damping system was designed to make automatically.

Among my increased responsibilities was maneuvering my spacecraft into proper position for firing the retro rockets. I would do so by aligning myself with the horizon by means of a line etched on my cabin window, aligning roll on the horizon, and yaw on a zero-yaw star—a star on my orbital path whose position didn't change. (I really had to know my stars, and considering that the only one I could point out when I came into the program was the north star, I had done a lot of astronomy homework.)

After flipping the switches on the control panel for firing the retro rockets, I'd have to be on the stick quickly to keep the spacecraft on the proper attitude to establish the correct trajectory for arriving at the targeted landing spot.

There was another concern about maintaining proper attitude: if I didn't hold the right position so that the multilayered fiberglass heat shield—several hundred pounds in weight—could take the brunt of the 3,000-degree heat that built up during reentry, my spacecraft and I would burn up during our descent into the Earth's atmosphere.

Even if we made a safe reentry, if the spacecraft kicked off horizontally to one side or the other and I was unable to correct its errant course, I could miss the recovery area by up to a hundred miles in latitude. If we were off vertically it could be worse: off by several hundred miles in longitude. In either case I'd land in the Pacific, the world's biggest ocean, but I sure didn't want to spend a lot of time sitting in a bobbing life raft waiting to be found.

When I passed over John Glenn's station on my second-to-last orbit, we reviewed the manual reentry procedures. There were a few things that John and I had disagreed on in the past, but when it came to flying and doing a job, John was cool and precise. In my situation, I appreciated both those attributes.

"We'll be ready the next time around, Gordo."

"Next time I see you," I responded, "I'll be firing."

The last orbit went by quickly, taken up with final reentry preparations.

At one point, Flight Director Chris Kraft came on the line to make sure I understood everything I needed to do and was ready. A former bigwig with NASA's predecessor—the National Advisory Commission for Aeronautics—Chris was a cool customer under the most adverse conditions. Customarily, Flight tended not to come on the radio himself but left the talking to CapCom. Hearing directly from the senior man in Mercury Control reinforced the level of their concern.

"I'm go here," I told Flight.

"Copy that," came his reply.

I wanted to get this done. For one thing, it was getting real warm. With no cooling system and without fresh air circulation, there was nothing to dissipate the heat from my body, and I was really sweating now. The temperature in my suit had risen to 110 degrees, and it was very stuffy. The cabin temperature was going up too, from the heat generated by the equipment that was still operating; it was nearly off the thermometer at 130 degrees.

More importantly, the carbon dioxide levels were continuing to climb inside the cabin and my suit. With the oxygen scrubbing system out, there was nothing I could do but ignore it and keep flying.

In training, we had run orientation tests to see if any of us were overly sensitive to carbon dioxide, but the levels we had worked with were not as high as what I was now experiencing. I was panting shallowly, a classic sign of carbon dioxide poisoning. At high enough levels, I would cease to function and eventually lose consciousness. Exactly when this would happen, the medical people could not calculate.

The ground updated me with the latest time sequence for retro-firing, and as soon as I was again within range of John's position, he began the final countdown for the retro sequence. It came through my headset loud and clear, reminding me of a starter's steady cadence in the moments before a track race, of which I had run many in my high school days in Oklahoma.

In addition to my Omega wristwatch, which had lost time during the heavy G loads I had experienced during liftoff, I was wearing an Accutron watch, which was still keeping accurate time. I followed the count on it, prepared to fire on my own if I lost contact with John.

"Five-four-three-two-one," John recited. "Fire!"

At 6:03 P.M., east of Shanghai, I pulled the toggle that activated the firing sequence for all three retro rockets. These rockets fired backward, or opposite the direction I was traveling in, to slow down my speeding spacecraft. Each fired for twenty seconds but overlapped—ten seconds after the first one ignited the second one went off, and ten seconds later the third one.

With the retros firing, the spacecraft was very unstable, and it started oscillating. With each oscillation, I had to be on the controls right away—correcting the roll, pitch, and yaw by manually operating push-pull rods connected to the attitude jets.

"How's your attitude, old boy?" John asked from far below.

"Looks like it's right on the old gazoo."

I had a good retro burn and was able to hold attitude throughout it pretty well, and afterward as the oscillations continued during reentry. As the heat increased, the Gs built up to 7 or 8 before slowly dropping off as my spacecraft decelerated.

Four minutes after retro-fire, I jettisoned the spent retropack—the rockets had done their job, slowing the spacecraft to about twelve thousand miles per hour, causing it to begin dropping out of orbit. The Earth's atmosphere would slow us down the rest of the way.

With the retropack gone, the heat shield was now fully exposed to provide its critical protection. I held the spacecraft at the zero-yaw 34-degree downward neutral angle with the heat shield leading the way. I had to be quick on the thruster jets to correct for the slightest oscillation, damping it out right away lest the divergence increase rapidly and throw the spacecraft into an out-of-control tumble. If that happened, I would lose my heat-shield–first attitude and the protection that it afforded against the 3,000-degree heat.

At 6:14 P.M. my spacecraft reentered the Earth's atmosphere and went into a communications blackout as intense heat forces and a sheath of ions built up on its surface and blocked out any electrical signals coming in or out. For several minutes I wasn't able to talk to anyone, and they weren't able to contact me.

During that time I continued to fight against oscillations to hold the right attitude.

I admit a devilish thought occurred to me during retro countdown: if I was four seconds late with the burn, I knew I'd land not off Midway, in the vicinity of the recovery fleet headed by the aircraft carrier USS *Kearsarge*, as planned, but right off Diamond Head in Hawaii, where I'd gone to college and lived for a number of fun-filled years, flying and surfing my way through the islands. I had visions of being greeted by a horde of ceremonial long boats filled with Aloha dancers and musicians strumming ukuleles. A colorful lei would be draped around my neck, followed by the traditional welcome kiss from a beautiful native girl in a shimmering grass skirt. Like a conquering hero I'd be led off to the nearest bar for a well-deserved Mai Tai. . . .

But any temptation passed because I knew I'd lose to Wally Schirra.

A career navy officer who took great pride in his service, Wally had explained to me that the correct way to come aboard an aircraft carrier, according to etiquette, was by the number three elevator just aft of the ship's island superstructure. Our challenge to each other was to see who could come closest to the recovery vessel. Wally had landed six miles from the carrier—the most accurate Mercury splashdown to date—and was quite pleased with himself. Now it was my turn. I took delight in knowing that Wally would be real burned if an air force flyboy landed closer to the recovery ship than he had.

Over lots of static, I finally heard something in my headset.

It was Scott Carpenter from a communications station in Hawaii.

"How—you doing?"

"Doing fine," I replied.

The drogue and main parachutes were on an automatic system designed to deploy at prescribed altitudes. But with no electrical power they were inoperable, so I went to manual override. At fifty thousand feet, I reached out to the upper left-hand corner of the instrument panel and pulled the toggle ring that controlled the drogue's release. I heard a loud pop and felt the jolt of the chute opening. The drogue was a funnel-shaped chute that looked almost like a windsock. It would not only help slow down the spacecraft but more importantly stabilize it and keep the right nose-up attitude prior to deployment of the main parachute. Without a drogue, the spacecraft could go into a spin, wrapping itself around the shroud lines of the main chute, which could collapse the chute and lead to a landing disaster.

At eleven thousand feet I pulled the toggle for the main chute and was gratified to feel it open with a big jolt that forced me into my seat. The spacecraft swung slowly under the huge canopy, ducking in and out of clouds as it floated gently toward the blue-green sea below and the carrier directly underneath.

Even with the huge chute, I was dropping at thirty-two feet per second, or approximately twenty-two miles per hour. Upon impact with the ocean, the force of the spacecraft drove it a good ten or twelve feet underwater before it popped back to the surface.

If it hadn't been for the wind causing drift, I believe I would have landed right square on the carrier deck. In fact, someone sent me a picture taken from the island of the carrier looking directly up at my spacecraft swinging under the chute.

When the spacecraft landed, a helicopter was immediately overhead, dropping navy frogmen to attach the flotation collar. Because I was so close to the recovery vessel, I elected to stay in the spacecraft rather than climb out and be hoisted to the copter like a sack of flour. A whaleboat came alongside and towed me the short distance to the carrier. Soon I was able to look out my window up at the big ship. It was a great sight. A boom came out from the deck and plucked the spacecraft from the water.

After splashdown, I had initially radioed the *Kearsarge* that I was in fine shape and apologized to the captain for not coming aboard by the number three elevator. The ship soon maneuvered into position so that the number three elevator could be used for my arrival.

Right after splashdown I had removed my helmet, and found the cabin stifling hot. When the hatch came off and I emerged from my scorched spacecraft, the first thing I did was take a big gulp of fresh sea air—then another. Man, it felt great.

Cheers arose from more than a thousand sailors on deck. A piece of red canvas had been rolled up to the hatch of my spacecraft; on either side were U.S. Marines in crisp dress blues standing at attention.

I found myself somewhat shaky on my feet after my six-hundred-thousand-mile journey and being weightless for thirty-two hours twenty minutes and thirty seconds (in spite of the difficulties, only a minute and a half longer than the flight plan called for).

I was greeted at the number three elevator by the captain, who saluted me crisply.

I saluted back, my arm feeling as heavy as a side of beef.

"Permission to come aboard, sir."

"Permission granted."

I was relieved, after the close call, to be back in one piece.

But mostly I was elated not only to have come aboard by way of the number three elevator but to have splashed down a mile closer to the USS *Kearsarge* than Wally.

My immediate future was filled with medical exams and debriefings.

They began on the *Kearsarge*, where a group of NASA people were awaiting my arrival—doctors, suit handlers, operations and planning personnel, and technicians who would accompany the spacecraft on its journey back to the Cape.

I spent the night on the carrier after enjoying some navy chow that tasted like the offerings of a five-star restaurant compared with the dried, concentrated stuff I'd been eating. The captain graciously moved out of his cabin, turning it over to me. As usual I had no problem sleeping as the ship steamed hard through the night for Hawaii.

The next morning I boarded a helicopter and headed for Honolulu. *That* ride I almost didn't survive. On the inbound flight I was asked to throw out a ceremonial wreath over the wreckage of the USS *Arizona* as a special Armed Forces Day tribute to the World War II dead. As we hovered above, I released the wreath, and in my still somewhat weakened state, nearly toppled out the door with it. Only a crewman's strong grasp saved me. I've often wondered how the navy would have explained the unceremonious loss of a returning astronaut over the Pearl Harbor memorial.

Much to my surprise, my nemesis, Walt Williams, was waiting for me in Hawaii on the tarmac with a firm handshake and congratulations. It was the first time the operations director had done that. "Gordo, you *were* the right guy for this mission," he said.

Who would ever have guessed that the only Project Mercury flight to return to Earth with manually controlled retro-fire and reentry would be mine? In Williams's own indomitable way, I think he was letting me know that he'd been glad, after all, to have a self-professed hotshot fighter jock at the controls.

Also waiting for me in Hawaii were Trudy and the girls, excited and anxious to hear all about the flight. After a parade in Honolulu, we returned to the Cape—the locals turned out to greet me as I stepped off the plane—and there was another parade in Cocoa Beach.

It was about then I learned that I'd been invited to give a speech before a joint session of Congress. But first came a big news conference. As I walked into the auditorium filled with reporters and camera crews, NASA's new deputy director for public affairs, Julian Scheer, came over and handed me a sheaf of papers.

"Here's the speech you're giving to Congress," Scheer said.

I eyeballed him. He was in his thirties, medium-sized, a snazzy dresser. In words and demeanor, he struck me as an aggressive PR type who always had an angle or two.

I flipped through the pages—there were about ten in all, typewritten.

I gave him back his speech.

"No, I'm not giving this speech," I said.

"What do you mean, you're *not*?" Scheer asked, horrified.

"I'm not giving a prepared speech."

"Oh *yes* you are."

"No, I'm not."

We went back and forth a few more times before I finally turned to some NASA officials nearby and said with clenched teeth: "If you don't get this son of a bitch out of here, I'm going to deck him right here in front of all the news media."

The PR guy went away, and avoided me for a week.

The press conference was nationally televised and lasted about an hour. I had gotten more comfortable with these gatherings since that first press conference when the Mercury astronauts were introduced to the world. I was the guy with all the answers that everyone came to hear. It was part debriefing and part performance, and no one could contradict me: I had been the only guy along for the ride.

I went over my experiences aboard *Faith* 7, proceeding step by step through the mission from countdown to recovery. I told about seeing John Glenn's "fireflies" and also a haze layer formerly mentioned by Wally Schirra during his *Sigma* 7 flight.

Asked how the sleep portion of the mission went, I grinned. "I can safely say I've answered with finality the question of whether sleep is possible in space flight."

Everyone laughed.

I said that one of the most surprising and astonishing features of the flight was my ability to distinguish objects on Earth from a hundred and fifty miles up. I told of seeing the African town where the flashing light experiment was conducted and of identifying several Australian cities, including the large oil refineries at Perth, also a train on a track in India, a big-rig truck rolling down a Texas highway, and a monastery high in the Himalayas.

Did I ever feel my life was in danger? I was asked.

I hesitated for a moment.

"I felt my options were rapidly decreasing," I said.

I didn't know how to explain it publicly, but I could not remember ever feeling that my life was in danger while flying. I had been in a tough situation where things went from bad to worse—but no, I had never thought I was close to buying the farm, even when, in retrospect, I probably was. Once, instead of ejecting, I rode down a jet fighter that was having electrical problems. It was in the dead of winter and I hadn't liked what I saw below: too much snow and ice. As I touched down on the runway, the plane lost all power and my controls went dead. The jet rolled out to a safe stop on the long runway, but had it happened a minute sooner . . .

A couple of days after the press conference, I was on a Gulf Stream jet flying up to Washington, D.C., with Deke Slayton, who was still not on active flight status but was working his tail off as head of the astronaut office, and NASA director Robert Gilruth, who was the biggest name in the manned space flight business as far as we astronauts were concerned because he had the final say on who went into space and when. We had our families along for the ceremonies.

During the flight, Deke came over and leaned down next to me. "Gordo, Bob is a little concerned," he said softly. "He's never heard you give a speech before."

It was true. Unlike some of the astronauts, I had always

shunned the public spotlight and speech-making whenever possible. When it came to public relations, the navy did a much better job of training and preparing its officers. The navy guys—Al, Wally, Scott, and John (the marines are part of the navy)—with their spit and polish and dress blues had a distinct advantage over us air force guys. Gus, Deke, and I came into the program looking like we were straight off the flight line, knocking the desert sand off our old leather jackets and boots. The fact that I had been the last to fly had aided me somewhat in my effort to keep as much of my privacy and anonymity intact as possible, although all those *Life* pictorial spreads didn't help. (I would end up on the cover of *Life* four times—twice in Mercury Seven group shots and twice alone, once before my Mercury mission and once afterward.) By the time the manned flights began, we were getting fourteen thousand pieces of mail a day from fans all over the world. An entire staff had been hired to handle the astronauts' mail operation.

"Bob's right," I agreed. "I've avoided giving speeches as much as possible."

"Well, Bob is just wondering: can you give a speech?"

Bob was shy and soft-spoken; I guess he wasn't comfortable telling me of his concerns. Instead he'd sent Deke over to lower the boom on me.

"Oh, I think I can probably manage."

I didn't mention that I had taken speech classes in college and felt fairly comfortable speaking to groups. Of course, this *was* a special group.

Deke wasn't satisfied. "Can you tell him what you're gonna say?"

I thought that was a reasonable request, considering who the audience would be. I took from my jacket pocket two 3 x 5 lined cards that I had scrawled notes on. Referring to them, I gave Deke a preview of what I had in mind to say.

"Let me go back and tell him. It'll make him feel a little better."

A few minutes later, Deke was at my side again. "Bob said

the speech sounds fine. It's going to be great." He handed me a sheet of paper with some writing on it. "We've gone through your onboard tapes and found this prayer you said in space. Bob asks that you finish your speech with it."

I looked down at the page and shrugged. "OK. I'll do that for him."

It had been a simple prayer and a spontaneous one, not meant for publication. I had said it not over the radio for the world to hear but into a small tape recorder as I flew over the Indian Ocean in the middle of the night on my seventeenth orbit. (Since it was difficult to take notes in space, I used the recorder to make running comments about various aspects of the flight.) It was subsequently labeled the "first prayer from space," although I'm sure that some travelers before me must have offered their blessings.

At least mine was the first prayer from space recited before Congress:

> Father, thank you, especially for letting me fly this flight. Thank you for the privilege of being able to be in this position; to be up in this wondrous place, seeing all these many startling, wonderful things that you have created.
>
> Help guide and direct all of us that we may shape our lives to be much better, trying to help one another and to work with one another. Help us to complete this mission successfully. Help us in our future space endeavors that we may show the world that a democracy really can compete, and that its people are able to do research, development, and can conduct many scientific and technical programs.
>
> Be with our families. Give them guidance and encouragement, and let them know that everything will be OK.
>
> We ask in Thy name.
>
> Amen.

"We admit we cringed when U.S. Astronaut Gordon Cooper began reading his flight prayer to the joint session of Congress," wrote *The New Republic,* a clip of which my mother pasted into my scrapbook. "The slow-spoken young man entered briskly, in a kind of lope, and the immense chamber turned pink with applauding hands. We looked down over his shoulder from the gallery; the self-possessed Oklahoman, with his hillbilly accent, was speaking from notes, without text."

It wasn't the first time I'd been told I talked like a hillbilly.

"In sophisticated society the polite thing is, if one prays, to keep such embarrassingly intimate matters to oneself. To dictate a prayer to a tape recorder over the Indian Ocean on the 17th orbit leads itself to caricature, but it was a simple and natural act for Major Cooper. Perhaps it revealed the source of composure that enabled him to take over the manual controls when automatic devices failed. His flight fell on the anniversary of Lindbergh's lonely trip to Paris, who carried with him, you remember, a letter of introduction to the Ambassador. Major Cooper, it occurred to us, carried with him a letter of introduction to God."

I was sorry my grandfather, who had died in 1954, hadn't lived to hear my prayer.

When I was ten years old, I had joined St. Paul's Methodist Church in my hometown of Shawnee, Oklahoma. At the time this had been no weighty decision for me to make or the result of any sudden revelation—it was just something I wanted to do. My grandfather on my mother's side, George Washington Herd, was a minister for the Church of Christ who traveled through the Indian territory—Arkansas, Texas, and Oklahoma—preaching the Gospel. He inspired me early on to read the Bible. Later, when I became an aeronautical engineer and astronaut, I saw no rift between science and religion. The more one learns about scientific endeavors, the more one must be open to the wonders of God's creation. The more I contemplated the complex workings of millions of planetary bodies and the unknown immensity of space, the more I realized what a fantastic miracle it all is.

I was later asked about the comments of a Soviet cosmonaut, Major Andrian G. Nikolayev, who said after his flight in *Vostok III* that he "didn't see God up there." My response was that if the cosmonaut had not known God here on Earth, he wasn't going to find Him a hundred and fifty miles up.

The day on Capitol Hill was exciting, with a parade down Pennsylvania Avenue, after which President Kennedy (who had called the USS *Kearsarge* so quickly to congratulate me after my return that I hadn't even had a chance to remove my space suit) pinned a medal on me in a ceremony in the White House Rose Garden.

President Kennedy had come to the Cape a few weeks before my flight to give me and the others at the Cape a little pat on the back. I'd given him a personal tour of my spacecraft and the launch facilities, then took a helicopter ride with him for an aerial inspection of the complete complex. He was the space program's biggest champion, and without him none of this would have been possible. President Kennedy was truly fascinated by the space program, *enthralled* with everything we were doing. I often thought that if he could trade places with anyone in the world, it would have been a Mercury astronaut.

"How do you feel?" he had asked me at one point from behind dark glasses in the brilliant Florida sunshine. "You up and ready?"

"I'm ready, sir," I said without hesitation.

He nodded thoughtfully. "You think you got all the training you need, Gordo?"

"*Yes, sir.*"

On a personal side, I found Jack Kennedy to be a terrific guy with a marvelous sense of humor. One time, sitting in the Oval Office with some NASA administrators, I was taking a pointed ribbing because some astronauts (myself included) had been seen by reporters in beach bars near the Cape with members of the opposite sex. It flew in the face of the NASA public relations machine, not to mention *Life* magazine's efforts to portray us as happily married family men. There had always been a conflict between trying to be the poster boys that NASA wanted us to be in public and being ourselves: red-blooded young men who had been indoctrinated in the rather carefree attitude toward life that most fighter pilots embrace. In light of the bad publicity that could result, I promised we'd be more careful. Observing this banter from his rocking chair nearby was President Kennedy. A few minutes later he got up, came over, and whispered in my ear, "You and I have the same problem."

"Major Cooper went further than any American in space," President Kennedy now told the gathering, making note of the Lindbergh anniversary. (Interestingly, our total elapsed flight times were only about five minutes apart.)

Of greater note, at least in my own family: I had returned on my mother's sixty-third birthday, presenting her, she told everyone, "with the best present ever." And here she was in the Rose Garden listening to the president of the United States.

Also very excited were both my grandmothers, who were well into their nineties. Grandma Cooper had arrived in Indian territory on horseback in the 1880s after her first husband had been killed by Indians, carrying her baby on her back papoose-style. She'd stopped at a remote trading post and asked for work, and the store's proprietor, Phillip Henry Cooper, hired her and later married her. Now she'd lived to see a descendant of hers *go into space*. What wondrous things mankind had accomplished in the course of one lifetime.

"Before the end of the decade," said President Kennedy, who had but six months to live, "we will see a man on the Moon; an American."

Before the celebration ended, there was a ticker-tape parade in New York—one of the biggest ever, I was informed, with the crowd estimated at four and a half million. My wife and I shared an open limousine with Vice President Lyndon Johnson for the whole route. We passed signs saying things like GORDO—YOU'RE SUPERDUPER! in block letters three feet high. Along the way, we set another record. It's a goofy stat, I know, but one I've never forgotten: 2,900 tons of ticker tape and confetti left on the street, the most trash after a New York City parade, according to the Sanitation Department. Eat your heart out, John Glenn.

There was a final "hometown" parade in Houston. By then I was exhausted from the arduous schedule of postflight parades, celebrations, and social activities. I had lost count of how many mayors and governors I had shaken hands with and how many awards I'd received. I told a lunch gathering the truth: "I believe the journeys we've had in the last few days have been longer than the flight I had in space—maybe not in mileage but in time and stress." I was ready to get back to work.

Less than a month after my mission, NASA administrator James E. Webb, testifying before the Senate Space Committee, said he felt that the energies and personnel of the country's manned space flight effort should now focus on the two-man Gemini and three-man Apollo programs—Gemini being a research and development program for Apollo, with Apollo achieving this country's goal of landing on the Moon.

Since we had accomplished most everything we set out to achieve several years earlier—specifically, determining that man and machinery were capable of making extended space flights and returning safely to Earth—NASA decided against another Mercury flight. This put us slightly ahead of schedule, perhaps, but with still an awful lot of work in front of us.

After four years, eight months, one week, and six successful space missions, Project Mercury, America's first manned flight program, came to a close.

"What a precision ending to Project Mercury, Gordo," Deke Slayton crowed, telling me that the technicians in Mercury Control swore I had pulled off the manual stick-and-rudder job with more precision than the autopilot had ever delivered.

"Aw, shucks," I said in my best hillbilly drawl. "We aim to please."

I would be the last American to fly alone in space.

But had I really been *alone*?

"GORDON COOPER'S UFOS"

Legend has it that as I circled Earth in my Mercury space-craft I became the first astronaut to see a UFO from space.

This tale has followed me around for years, repeated in numerous articles, books, and television documentaries. It is still a topic of discussion for serious UFOlogists, as well as active chat groups on the Internet. I was recently sent a sheaf of essays and discussions headlined "In Search of Gordon Cooper's UFOs" that had been printed off a popular Internet site and that included this story at length.

Here is how it is *reported* to have happened:

"The object which approached Major Cooper was also seen by the two hundred people at the Muchea Tracking Station near Perth, Australia," according to one published account. "It was reported twice on the NBC radio network before Cooper had been picked up by the recovery vessels. He was not permitted to comment on it."

Another account:

"Major Cooper, on his final orbit over Australia, contacted

the tracking station in Muchea and reported that a greenish object, moving east to west, was approaching his capsule. The object was tracked by Earth-based equipment in Perth."

The story spread and grew more colorful as the years passed. A 1973 book, *Edge of Reality,* claimed: "Gordon Cooper reported a greenish UFO with a red tail during his last orbit. He also reported other mysterious sightings over South America and Australia. The object he sighted over Perth, Australia, was caught on screens by ground tracking stations."

The only problem with these stories: *never happened.*

I saw *no* UFOs from space.

I made *no* radio transmissions about any object approaching my spacecraft, and have the onboard tapes of the flight to prove it.

No eyewitnesses from Perth have ever stepped forward, and NBC radio has reported finding "no tape or transcripts concerning any such incident."

I have publicly denied this story again and again, but it won't go away.

Similar purported UFO sightings by other astronauts, including Wally Schirra (*Mercury 8*), Jim Lovell and Frank Borman (*Gemini 7*), and Neil Armstrong and Buzz Aldrin (*Apollo 11*), never happened either, nor were there sightings or pictures taken by astronauts of UFOs or alien structures on the Moon, as has been claimed in some quarters.

It got so bad that there were deliberately falsified tapes of communications with astronauts, where UFO material was simply edited in. Falsified reports and general misinformation on the subject do a great disservice to us all. I know credible people, including military and airline pilots, who have had legitimate sightings but haven't filed reports for fear of being grouped with all the nuts who don't care about the truth.

One friend is a veteran captain for U.S. Air. He has had four real good sightings during his career, one right off his wing, but was warned by management not to discuss or report UFO sightings because they would be "bad for business."

The military isn't any more forthcoming. Two air force test pilots, close friends of mine, were returning from a conference at Wright-Patterson in 1958, flying a T–33 jet trainer at thirty thousand feet. Along about Amarillo, they were contacted by Albuquerque Regional Flight Control, which monitors air traffic for a large southwestern section of the United States, and asked if they could see an aircraft in front of them. They reported something glinting in the sun—too far ahead to identify. Flight Control said they hadn't been able to establish communications with the craft, which radar showed was "going right along our airway," and asked if the T–33 could get closer for a look-see. They bent the throttle to the stop, and closed on the object. The first thing they noticed was that it wasn't leaving a contrail—no evidence of exhaust—even though they looked back and saw that they were putting out a big contrail at the same altitude. They saw that the craft didn't have wings. Then they pulled right next to it, and saw it was a big metallic saucer.

The jet and saucer stayed together for about ten minutes, flying in close formation. Then the saucer tipped up, and streaked up and out of sight. When they landed, the air force pilots went into an office and filled out a report about the UFO encounter, which in this instance was documented by radar reports. There was no investigation, and my friends never heard anything more about the incident.

I know of only one possible UFO seen in space by an astronaut, which was duly reported at the time. It occurred in 1965, when James McDivitt and Ed White were passing over Hawaii in *Gemini 4* and spotted what McDivitt described as a "weird-looking metallic object" with an "arm sticking out." The object was moving away from the spacecraft. It wasn't on their list of space junk, so he took some pictures. It was so bright with the sun shining off it that it was difficult to see much detail, and the astronauts frankly didn't know what they were observing. Later there was some speculation that it could have been the second stage of their Titan II booster, but McDivitt had seen the rocket section trailing them earlier and identified it as such.

To my knowledge, that was the only unidentified sighting by astronauts on any Mercury, Gemini, or Apollo missions, despite all the stories that have circulated.

During the decade that I was in the astronaut corps, NASA did not brief us on UFOs—nothing about what to do or not do if we had a sighting in space. Even without official guidance, I think any of us, being experienced pilots, would have done what Jim McDivitt did: observe, try to document, and report.

NASA and the space program were naturally drawn into the UFO phenomenon. It made sense too. As humans explored space, it was reasonable to imagine that other beings in the universe were doing the same. And if other civilizations were visiting us, it stood to reason that the astronauts would see something in space. Or perhaps the visitors would want to check out our space travelers. Then again, if a civilization was advanced enough to reach Earth from even the nearest large galaxy, Andromeda, some two million light-years away, it probably could have kept an eye on our first baby steps into space without our even knowing it. In the great scheme of things, our manned space travel has been pretty insignificant thus far.

In 1984 NASA founded the radio astronomy project called Search for Extraterrestrial Intelligence (SETI) to scan the skies for signals from deep space. NASA continued to fund SETI until 1993, when Congress ended funding. Since then, financial support has come from a number of computer pioneers, including William Hewlett, the late David Packard, and Microsoft cofounder Paul Allen. SETI, which was the basis for the popular movie *Contact*, continues to monitor a wide range of microwave radio frequencies on sixty huge dishes around the world, with professional and amateur radio astronomers listening for any possible signals from an extraterrestrial civilization. Everyone who works in the program has a twenty-four-hour beeper and will be notified if and when an alien signal is picked up. "Most of us are believers," says Dr. Peter Backus, a Ph.D. radio astronomer and associate project manager of SETI, who has worked in the field for seventeen years, "and think it will happen in our careers." SETI's current annual budget is four million dollars, and it employs eighteen people full-time.

Although NASA won't officially confirm the existence of UFOs or even suggest the possibility of extraterrestrial intelligence, there must have been a consensus among its top administrators that *something* was out there. If not, why would they have spent some sixty million dollars on SETI over a decade?

The history of life on Earth suggests that life could develop elsewhere given a suitable environment and sufficient time. We know from what we can see and measure that there are at least four hundred thousand other planets that could have identical or reasonably similar atmosphere, temperature, and gravity to what we have on Earth. If you start plotting three-dimensional galaxies, ours sits out on the fringe of this big accrual of galaxies. And we haven't yet seen beyond those galaxies, so who knows what may be beyond *them*? I can't believe that God would populate only one planet way out here in the hinterland.

Ever since I looked up at the stars as a boy, I've felt that there had to be some interesting forms of life out in space for us to discover and get acquainted with. And I had the same feelings when I was in space—looking *out there*, toward deep dark space. Any explorer or researcher must have a certain amount of curiosity in his makeup that drives him to discover things not yet found. I have always had that curiosity, and it was a big reason why I wanted to become an astronaut.

I don't believe in fairy tales, but when I got into flying and military aviation, I heard other pilots describe too many unexplained examples of UFOs sighted around Earth to rule out the possibility that some forms of life exist beyond our own world. I had no evidence at the time that these examples conclusively proved anything, but the fact that so many experienced pilots reported strange sights that could not easily be explained only heightened my curiosity about space.

And then—I had my *own* UFO sightings.

Air Forces Europe was my first assignment.

I was assigned to the 525th Fighter Bomber Squadron, one of the air force's first operational jet squadrons, stationed at Neubiberg Air Force Base in West Germany. The year was 1950; I was a twenty-three-year-old second lieutenant.

We flew patrol along the borders of Communist East Germany, Czechoslovakia, and Poland and found ourselves up against superior MIG–15s in slower and less maneuverable F–84 Thunderjets.

The F–84s had serious problems—we were blowing about fifty-five engines a month in our group of seventy-five aircraft. We all got a lot of dead-stick landing experience. For some, the experience proved fatal: we lost twenty-one pilots in two years.

The MIG–15 had a higher top speed (six hundred and fifty miles per hour), a higher rate of climb, and a higher ceiling (fifty thousand feet) than our fighter, and with experienced pilots at the controls, they strayed across the border looking to mix it up. One morning a flight of MIGs buzzed the center of Munich, well inside the U.S. sector of West Germany.

Things got pretty tense at times. While the U.S. State Department told the world that our fighters were flying "unarmed" border patrols, both sides went up with charged guns and live ammunition. Planes on both sides often came back with holes in the fuselage. One U.S. jet was even shot down, although the pilot parachuted to safety.

It was only after we received new F–86 Sabrejets that the MIGs met their match and stayed on their side of the fence.

While in Germany I became an experienced pilot accustomed to flying in all weather conditions and at night. For more than two years I was going to school three nights a week at the University of Munich, earning credits toward my bachelor's degree. I would hop in a fighter and fly five hundred miles to class, then return to the base around midnight.

It was in Europe, in 1951, that I saw my first UFO.

When the alert sounded, my squadron mates and I dashed from the ready room and scrambled skyward in our F–86s to intercept the bogies.

We reached our maximum ceiling of around forty-five thousand feet, and they were still way above us, and traveling much faster. I could see that they weren't balloons or MIGs or like *any* aircraft I had seen before. They were metallic silver and saucer-shaped. We couldn't get close enough to form any idea of their size; they were just too high.

For the next two or three days the saucers passed over the base daily. Sometimes they appeared in groups of four, other times as many as sixteen. They could outmaneuver and outflank us seemingly at will. They moved at varying speeds—sometimes very fast, sometimes slow—and other times they would come to a *dead stop* as we zoomed past underneath them. We had no idea whether they were looking at us or what they were doing. They came right over the air base at regular intervals all day long, generally heading east to west over central Europe.

I suppose there were reports filed by officers a lot more senior than I—still a lowly second lieutenant. But as far as I know there was no official investigation.

Since the UFOs were too high and too fast for us to intercept, we eventually stopped going up after them. Through binoculars we looked to the sky in awe at these speedy saucers. Our worst fears were that the Soviet Union had developed something for which we had no match. And if they weren't from anywhere here on Earth, we wondered aloud—*Where did they come from?*

• • •

After Germany, I attended the Air Force Institute of Technology at Wright-Patterson Air Force Base in Ohio for two years, determined to finally earn my degree. I had trouble at first getting my study habits down, but midway through the course work I became interested in rocket propulsion and air-craft design. An interested student makes for a good student, and in 1956 I graduated on the honors list with a Bachelor of Science degree in aeronautical engineering.

Following my graduation I was selected for the Air Force Experimental Flight Test School at Edwards Air Force Base in the California desert. Test pilot school was for the best of the best among air force pilots, and after graduation I was assigned to the Fighter Section of the Experimental Flight Test Engineering Division at Edwards as test pilot and project manager.

On May 3, 1957, I was a captain and had a crew out film-ing an Askania-camera precision landing system we had installed on the dry lake bed. The Askania automatic system took pictures—one frame per second—as a plane landed to measure its landing characteristics. The two cameramen, James Bittick and Jack Gettys, arrived at Askania number four site a little before 8 A.M., armed with still and motion picture cam-eras.

Later that morning they came running in to tell me that a "strange-looking saucer" had come right over them.

"It didn't make any noise at all, sir," one of them said.

"Not a sound," the other one agreed.

I knew these enlisted men to be old pros, but they were really *worked up*—excited and frightened in the same breath. They were accustomed to seeing America's top-performance experimental aircraft taking off, screaming low overhead, and landing in front of them on a daily basis. Obviously what they had seen out on the dry lake bed was something quite different, and it had unnerved them both.

They told me they had just about finished their work when the saucer flew over them, hovered above the ground, extended three landing gear, then set down about fifty yards away. They described the saucer as metallic silver in color and shaped somewhat like an inverted plate.

I had heard insider reports from people I trusted about a mysterious crash near Roswell Air Force Base in New Mexico in 1947. (Roswell was the home of the air force's 509th Bomb Group, which at the time was the nation's only nuclear strike force—a fact that was a closely held secret. The *Enola Gay,* the plane that dropped the atom bomb on Hiroshima, was based at Roswell.) On July 7, crash debris was found in a rancher's field. The following day the Roswell base commander, Colonel William Blanchard, released a story to the news media that the air force had recovered a "flying saucer." It ran as front-page news in the *Roswell Daily Record* with an eye-popping front-page headline: AIR FORCE CAPTURES FLYING SAUCER ON RANCH IN ROSWELL. Within days, the official story changed—it wasn't a flying saucer after all, explained Brigadier General Roger Ramey, commander of the 8th Air Force, headquartered in Fort Worth, Texas, but only a "crashed weather balloon." A pilot and air force major who was a good friend of mine had been at Roswell. He had seen some of the debris recovered at the crash site, and he told me it sure wasn't a weather balloon. Although he had to be careful about what he said due to stringent security around the entire incident, he made it clear that what crashed that day was an aircraft of some type, and that bodies of the crew were recovered.

As late as 1994, when the congressional General Accounting Office requested information about what has since become known as the "Roswell incident," perhaps the most famous of all UFO cases in the world, the air force stuck to its weather balloon story. "An exhaustive search for records found absolutely no indication what happened near Roswell in 1947 involved any type of extraterrestrial spacecraft," an air force report concluded. As for persistent reports—from military as well as civilian eyewitnesses—that alien bodies were observed at the crash site and later at the base, the air force suggested they were "anthropomorphic test dummies that were carried aloft by high-altitude balloons for scientific research."

I knew an air force master sergeant assigned to a team that received an emergency call-out from Washington, D.C., to the Pacific southwest (not Roswell). He told me they reached a canyon and found a wreckage site. According to this friend—and I had been around him enough to consider him a reliable guy— a metallic disk-shaped vehicle had crashed, and sitting atop the wreckage were two very human-looking fellows in flight suits, waving at them. They were hustled away, and the sergeant never found out who they were or what happened to them.

After my own UFO experiences in Europe, I was not about to discount *any* of these stories, especially coming from people I had served with and trusted.

These two cameramen were trained photographers and had cameras and film with them. I quickly asked the obvious question: "Did you get any pictures?"

"Oh yes, sir. We were shooting the entire time."

They said they had shot images with 35-mm and 4-by-5 still cameras, as well as motion picture film. When they had tried to approach the saucer to get a closer shot, they said it lifted up, retracted its gear, and climbed straight out of sight at a rapid rate of speed—again with no sound. They estimated the craft to be about thirty feet across. It had a silver color to it and seemed to glow with its own luminosity.

I told them to get the film to the lab right away.

I had to look up the regulations to see how I should report the incident. There was a special Pentagon number to call in the event of unusual sightings. I called it and started with a captain, telling him we'd just had a sighting and landing of a "strange vehicle that didn't have wings on it." The captain quickly passed me to a colonel. Eventually I was talking to a general, repeating for the third time what had happened that morning. He ordered me to have the film developed right away but "don't run any prints" and to place the negatives in a locked courier pouch to be sent to Washington immediately on the base commanding general's plane.

I wasn't about to defy the Pentagon general's order about no prints—a surefire way to end my career or, at the very least, lose my top-secret clearance and my test pilot job. But since nothing was said about not *looking* at the negatives before sending them east, that's what I did when they came back from the lab.

I was amazed at what I saw. The quality was excellent, everything in focus as one would expect from trained photographers. The object, shown close up, was a classic saucer, shiny silver and smooth—just as the cameramen had reported.

I never saw the motion picture film. Before the day ended, all the negatives and movie film had left on the priority flight for Washington.

Considering what the men had seen, and particularly the photographic evidence they had brought back with them of a UFO touching down on Earth, I expected to get an urgent follow-up call from Washington, or the imminent arrival of high-level investigators. After all, a craft of unknown origin had just overflown and landed at a highly classified military installation.

Strangely, there was no word from Washington, and no inquiry was launched. Everything was kept under wraps, as if the incident never happened. Through the years, it would have been easy for me to forget the entire matter—*if I hadn't seen those photographs*.

The incident report was supposed to wind up at Wright-Patterson Air Force Base, home of the air force's official UFO investigation, Project Blue Book. I don't know who saw the photographic evidence or what happened to the photos once they were printed. All I know for sure is that the evidence I'd seen with my own eyes vanished. After putting the negatives and film on the plane to Washington, that was the last I heard or saw of them.

Except that two years ago I was contacted by an independent researcher who said he'd tried to uncover information about the pictures of the Edwards sightings through the Freedom of Information Act. He said he'd found a reference in an old Blue Book report of pictures having been taken of "something unusual" at Edwards, but that was it.

Between 1948 and 1969, Project Blue Book investigated 12,618 reported UFO sightings. Of these, 11,917 were dismissed as balloons, satellites, aircraft, lightning, reflections, astronomical objects such as stars or planets, or outright hoaxes. The remaining 701 were classified as *unexplained sightings,* not UFOs.

Among Project Blue Book's 701 "unexplained sightings":

- On March 13, 1951, at McClellan Air Fore Base, U.S. Air Force First Lieutenant B. J. Hastie saw a cylinder with twin tails, two hundred feet long and ninety feet wide, turn north and flew away at incredible speeds. Length of sighting: two minutes.

- On March 24, 1952, at Point Concepcion, California, a B–29 navigator and radar operator tracked a target for twenty to thirty seconds traveling at an estimated three thousand miles per hour, about four times faster than the speed of sound. (The X–2 had exceeded the speed of sound only three years earlier, and there was nothing faster in our aviation arsenal.)

- On January 10, 1953, in Sonoma, California, retired Colonel Robert McNab and an employee of the Federal Security Agency saw a flat object make three 360-degree right turns in nine seconds, make abrupt 90-degree turns to the right and left, stop, and accelerate to high speeds before flying out of sight. Length of sighting: seventy-five seconds. (Such movements were not—and are not—possible with conventional aircraft. They are beyond the tolerance of our airframes and far exceed the G-load tolerance of a human pilot. I've thought about how these maneuvers might be possible, and all I can come up with is that these vehicles have somehow created an artificial gravity within the craft itself that frees them from the gravitational forces that pin the rest of us to Earth.)

- On February 2, 1955, at Miramar Naval Air Station, California, U.S. Navy Commander J. L. Ingersoll said that a highly polished sphere, reddish-brown in color, seemed to fall from the sky, then instantly accelerate to fifteen hundred miles per hour.

One famous sighting that was also dismissed by the air force became known as the "Washington merry-go-round." On July 15, 1951, a number of objects appeared over D.C., seen by a vast number of witnesses. They were tracked by radar—seven blips in one corner of the screen, heading for the White House. Jets scrambled, and the blips *disappeared*. The jets returned to base, and the UFOs *reappeared*.

UFOs appearing with impunity over the heart of the U.S. government and military establishment represented a grave embarrassment to the Pentagon. At a military press conference about the event—the biggest air force press conference since World War II—General John Sanford said that they were not secret aircraft in the U.S. arsenal. "A certain number of credible witnesses are seeing incredible things," he said. But soon the lid slammed shut and the official explanation was put forth for the Washington sightings: "temperature inversion." The temperature aloft normally drops 1 degree for every two thousand feet.

An inversion is when the temperature remains constant or goes up with higher altitude. The problem with using temperature inversions as an official explanation for UFO sightings, as has been done much too frequently when there is no other explanation, is that inversions are not known to cause "hard" radar returns.

One of the most puzzling and unusual cases in the "radar-visual" category took place on the night of August 13, 1956, when radar operators at two military bases in the east of England repeatedly tracked single and multiple objects that displayed high speed as well as rapid changes of speed and direction. Two jet interceptors were sent up and were able to see and track the UFOs. At one point the speed of the UFOs was estimated at over twelve thousand miles per hour. One UFO then slowed and got on the tail of a jet fighter that could not shake it. The UFO had the capability to stop suddenly and make sudden course changes without slowing down. In a 1969 air force–funded UFO study at the University of Colorado under Dr. Edward Condon, this sighting was singled out due to the "rational, intelligent behavior of the UFOs . . . [suggesting] a mechanical device of unknown origin as the most probable explanation."

When Project Blue Book was terminated by the air force in 1969, it left the world with three conclusions:

- No UFO reported, investigated, and evaluated by the air force has ever given an indication of threat to our national security.

- There has been no evidence submitted to or discovered by the air force that sightings categorized as "unidentified" represent technological developments or principles beyond the range of present-day scientific knowledge.

- There has been no evidence indicating that sightings categorized as "unidentified" are extraterrestrial vehicles.

Playing at the Shawnee airport, age eight. By then I was flying with the help of blocks Dad built for the pedals and lots of cushions. I had taken my first flight at age five in a Curtis Robin high-wing monoplane, shown in background.

My high school football team played for the state championship my senior year with me at halfback. I turned down the possibility of a football scholarship to Oklahoma A&M to enlist in the Marine Corps, hoping to do my part in World War II as a rifleman.

At age sixteen, with a camera in my hand. I always had a camera close by and would be the first astronaut to take an assortment of photographs from space—including some that the military and even the president of the United States were none too happy about.

Being sworn in as the newest lieutenant in the U.S. Air Force by my father, Lieutenant Colonel L. Gordon Cooper, Sr., September 1949.

My fighter squadron in Germany 1951, about the time that I chased UFOs over the skies of central Europe. I'm in the first row, fourth from right. Not many of us in this picture are left; most died long ago in plane crashes.

Descending the stairs at Hangar S in full flight suit
prior to my Mercury mission.

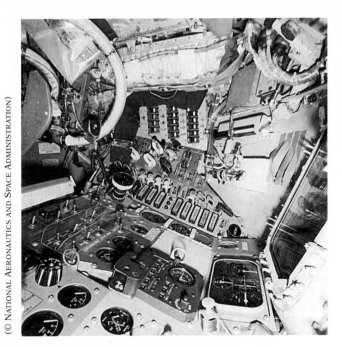

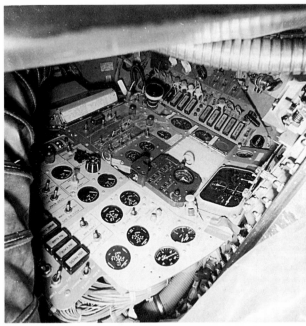

The interior of my Mercury spacecraft, *Faith 7,* which stood nine and a half feet tall and measured six feet across the base. By the time the spacecraft was filled with instruments and controls, it was like going into space while sitting in a small bathtub.

The candle was lit. Over my headset came the excited announcement: "We have liftoff!" The weight of my Atlas rocket on the launch pad was 260,000 pounds—fully 200,000 pounds of that in rocket fuel alone.

Happy and exhausted following my thirty-four-hour Mercury mission, right after the hatch was blown aboard the carrier USS *Kearsarge* southeast of Midway Island in the Pacific.

Ceremonies at the White House. My mother, Hattie, is far left; my wife, Trudy, next to her; then my two daughters, Camala and Janita. My buddy Gus Grissom can be seen between me and JFK. To the right of the president: Al Shepard, Wally Schirra, and Scott Carpenter.

Speaking before a joint session of Congress after my Mercury mission. "The self-possessed Oklahoman, with his hillbilly accent, was speaking from notes, without text," reported *The New Republic*.

My parade set a record: 2,900 tons of ticker tape and confetti
left on the street, the most trash after a New York City parade,
according to the Department of Sanitation.

It was a perfect moment, when time seemed to stop. I was seeing our planet from space for the first time, and I wanted to live within the moment for as long as possible.

The high plateau area near the Himalayas, taken while I was supposed to be sleeping. This image was shot with a 70-mm modified Hasselblad camera equipped with an 80-mm F:28 lens. Ansco color ultraspeed ASA 160 film was used.

No, this isn't a Middle East peace delegation. It's just us, the Mercury 7 gang, following a few days of desert survival training in Nevada—in case we ever missed our ocean landing zone and had to fend for ourselves. From left: me, Scott Carpenter, John Glenn, Al Shepard, Gus Grissom, Wally Schirra, and Deke Slayton.

Notwithstanding my smile, our flight suits were not real comfortable.

"Excuse me, fellows. Isn't there a better way to get aboard ship?" That's me being hoisted up to a navy helicopter during recovery operations in the Atlantic following our *Gemini 5* mission.

Atop a Titan rocket, back to space aboard *Gemini 5*
for an eight-day mission.

My *Gemini 5* crewmate Pete Conrad, left, with me at pad 19 shortly before our eight-day mission.

The Mercury Seven astronauts in flight gear, next to a jet. *Left to right:*
Lt. M. Scott Carpenter, USN; Capt. L. Gordon Cooper, Jr., USAF;
Lt. Col. John H. Glenn, Jr., USMC; Capt. Virgil I. "Gus" Grissom, USAF;
Lt. Comdr. Walter M. Schirra, Jr., USN; Lt. Comdr. Alan B. Shepard,
Jr., USN; and Capt. Donald K. "Deke" Slayton, USAF.

In other words, an official whitewash of the entire UFO matter.

Around the time I was chasing UFOs in a jet over Europe, hundreds of other pilots and military personnel were having similar experiences. From 1947 through 1951, according to Project Blue Book records, the air force logged seven hundred UFO reports, an average of just over a hundred and fifty per year. The first serious wave of UFO sightings began in 1952, within months of my own UFO experience. By the first half of that year, there were more than *four hundred reports,* many of them coming from military jet pilots sent aloft in response to radar or visual sightings from the ground. And according to the air force there was *no evidence* of UFOs?

The first American president to answer a question at an official press conference about UFOs, following a rash of public sightings, President Harry Truman said on April 4, 1950: "I can assure you that flying saucers, given that they exist, are not constructed by any power on Earth."

It's clear that our government started out in the 1940s trying to keep UFO information quiet because of concern about public panic at the thought of vehicles from space being able to outperform our best aircraft by multitudes, which meant that we would have little defense in the case of interstellar warfare. That concern was expressed by none other than U.S. Army General Douglas MacArthur, who warned, in his last major address to Congress in 1955, that the people on Earth must unite to "make a common front against attack by people from other planets."

I give the public more credit than our government has at times. Most people want to know what's going on in the world around them and would rather hear the truth, whatever it is, than a pack of lies. After the government told the first untruth about UFOs, it had to tell another to cover that one, then another, and another. It just snowballed. And right now I'm convinced that a lot of very embarrassed government officials are sitting in Washington trying to figure a way to bring the truth out. They know it's got to come out one day, and I'm sure it will. America has a right to know.

But as I see it, our government is now trapped in a big box of old lies. It's going to take a lot of courage on the part of some future administration to say, "Folks, our government has been lying to you all these years. Now we're going to come clean and tell you the real truth." As I said, that's going to take *courage,* something there doesn't seem to be a surplus of in Washington these days.

A recent *Newsweek* poll found that 48 percent of Americans believe that UFOs exist, with the same number thinking there's a government plot to cover up the whole thing; 29 percent believe we've made contact with aliens; and 10 percent—some twenty-six million Americans—say they have seen a UFO in their lifetime.

Does that sound like a public not ready to hear the truth?

There are nonbelievers, of course. Considering all the untruths and disinformation that has surfaced on the subject, I can hardly blame most of them. As for those individuals who are afraid of the unknown, it has always been my belief that the more we know about anything, the less fearful we are of it.

The strong photographic-evidence case from Edwards was not included on the list of Project Blue Book's "unexplained sightings." As far as the military was concerned, the Edwards incident was stamped *Case Closed.*

I would later hear that the two cameramen were advised by higher-ups that the object they saw and photographed was a "weather balloon distorted by the desert atmospheric effects." (From *fifty yards?*)

There was nothing more the cameramen could do—they no longer even had their film—and nothing I could do either. I was a junior officer holding a high-level security clearance to fly some of this country's top aircraft.

Although I didn't know it at the time, news of the sighting ran in numerous newspapers and even made national wire services. A month after the incident, Major Robert Spence, of the Office of Information Services at Edwards, wrote to an inquiring news reporter:

> The alleged UFO was conclusively identified as a balloon from a weather unit a few miles west of the observers' location. This was corroborated by an independent report which discloses that this balloon was being tracked at that time with precise recording devices. . . . These are typical of a number of similar reports received by the Air Force, which upon investigation were found to be balloons, the odd appearances being caused by reflection of sunlight. It is the opinion of the Air Force that any attempt to attribute anything unusual or mysterious to the incident is unwarranted.

Nobody tried that weather balloon line on me. If they had, I would have told them what I thought: *I've never seen a saucer-shaped balloon with three landing gear on it.*

5
AMELIA, PANCHO, AND DAD

I took my first flight when I was five years old.

Flying has been in my blood ever since. I have my father, Leroy Gordon Cooper, Sr., after whom I was named, to thank for my love of flying.

During World War I, Dad ran away at fifteen to join the navy. When they discovered his age a few months later, they gave him an honorable discharge. A year later he enlisted again. This time his real age wasn't found out until he was seventeen. He served as a bugler aboard the presidential yacht, *Mayflower*, which cruised up and down the Potomac with President Wilson and his guests. After the war, Dad became a member of the Army National Guard, flying the JN–4 DeHavilland biplane, known as the "Jenny," which had been used for training pilots during the war. He never had any formal military flight training but became an excellent pilot.

Both my dad and my mother—Hattie Lee Herd—hailed from Maud, an oil-boom town of three thousand residents some twenty miles from Shawnee (population eighteen thousand).

They had known each other all their lives. When he returned home after the war to finish high school, she was his teacher, although she was only two years his senior. Mother was a very pretty, fair-complected redhead with long hair and blue eyes. They soon fell in love and married. Their only child, I came along about five years later.

After graduating from high school, Dad went on to college, completing his four-year degree in two years. Then he went right to law school. After earning his degree in an accelerated program for returning vets, he practiced law before becoming a district judge. Soon after World War II was declared, he went on active duty with the Army Air Corps—which predated the air force—as legal staff and flight crew in the South Pacific. When the Judge Advocate General's office was formed, his legal background was in demand and he went with JAG, which gave him the opportunity to argue cases before the Supreme Court. He made a career of the service, retiring as an air force colonel.

My first flight was in a single-engine Curtis Robin high-wing monoplane, sitting on my father's lap. I loved it from the moment Dad cranked the engine and I heard the low-throated rumble of the engine as it shook the light fuselage. We took off on one of the grass fields at Regan Airport, just outside of Shawnee, and flew around. No sooner had we landed than I was asking Dad when we could go up again.

Ours was a comfortable life. In addition to all the necessities of life, we always owned our own home and our own plane. When I was nine years old, my father presented me with a lawn mower. "Here's how you earn your luxuries," he said. By the time I was twelve, I had my own checking and savings accounts.

We kept a family plane at Regan field for nearly as long as I can remember. Dad bought a Commandaire biplane, which had been designed as a World War I fighter, although only about a hundred were built. A three-seater with two open cockpits, it was a powerful little bird that could almost do a loop on takeoff.

Dad flew from the rear seat, where the main instruments and controls were located. Mom and I sat in the forward cockpit, which had its own stick and rudder pedals. (Mom flew with me well into her eighties. Whenever I showed up in a plane to visit her, she'd come out and hop in for a ride.)

By the time I was eight, with the help of blocks Dad built for the rudder pedals and lots of cushions so I could see out the cockpit window, I was allowed to take over the controls from the front seat with Dad right behind me in the main cockpit. We always flew whenever we visited my uncle in Los Angeles or went to our cabin on government-owned forest land high in the Colorado mountains.

Our plane had a range of a hundred and fifty to two hundred miles. We'd take off and follow the highways, most of them gravel in those days. When we needed gas, we did what was common practice among pilots then: kept an eye open for a gas station. When one came along, we'd land on the road, taxi up to the pump, and say, "Fill 'er up." Then we'd zoom off.

There weren't as many regulations governing civilian aviation back then, and those on the books were not always enforced. I was familiar with various flying maneuvers and knew how to handle an aircraft long before I was in my teens.

The head of the Federal Aviation Agency (FAA) in that region of Oklahoma was a close friend of my father's (my mother and father seemed to know just about everybody in the area). He knew I was flying without a license—I was a long way from sixteen, the mandatory age to get a license—and never said a thing. By the time I was twelve I was flying solo, even though I had not yet had any formal lessons.

I did get in trouble once with the FAA guy when I was about fifteen. I was at the controls of a J–3 Cub, dropping down over a row of big trees, rolling my wheels down my aunt Jewel's lawn, and pulling up before I hit a row of big trees on the other side. I did this several times, while waving to my aunt and cousins, who were enthusiastically waving back. The FAA inspector happened to be driving down the highway and saw this airplane doing crazy stunts. He went like gangbusters out to the airport and was waiting there when I came in.

"I'm gonna take you out by the hangar and give you a good thrashing!"

I crossed my heart not to do it again. After much pleading on my part, he came around to letting me off, warning that if he caught me doing another dumb stunt like that, "There'll be hell to pay, young man."

He never caught me doing anything like that again, let me tell you. Of course, by then I knew what kind of car he drove, and I kept a *very close* lookout for that big black Packard.

I spent a lot of time at the local airport, washing and polishing aircraft and helping out with odd jobs just to be around the men and machines that took to the air. I jumped at every opportunity for a flight. The latter half of my junior year in high school, I managed to get into a formal flight-training program. Also, my high school offered several courses in aeronautics, and I took every one of them. Dad told me that flying lessons were the one luxury he would help me finance, although I earned some of the money by my work at the airport. After school I'd ride my bike over to the airport for lessons and solo practice. Occasionally I'd cut classes to fly. Luckily, school came pretty easy for me, and I never had to work real hard at my studies.

Periodically my parents would take me on long trips during the school year, sometimes as far away as Mexico. We always took along my schoolbooks and a schedule of classroom assignments so I would not fall behind, and my mother was my tutor. She enjoyed the outdoors, cooking on an open fire and curling up in a sleeping bag under the stars, as much as my father and I did. As a result, we did lots of camping as a family. We occasionally took packhorses up into the mountains of Colorado.

Mother was an avid reader and a born teacher, although she didn't return to work after I was born. She focused her attention on raising me and managing the household, both of which she did with great skill. While Dad taught me how to fly, hunt, and fish, my mother taught me the importance of learning—and, much to my delight, how to drive when I was eleven (after a couple of years of sitting on her lap behind the wheel). She liked to fish too, and caught her share. Other times she'd leave the fishing to us guys and would head up a steep rock slide to pick wild raspberries at the top for her jam.

Basic to my parents' philosophy was the idea that there was a great deal to education other than books. This does not imply any neglect of my classroom education—after a tornado blew off the top floor of my elementary school, my parents were most insistent that I complete my assigned homework that night. They simply appreciated the other things to be learned from travel and meeting people.

Early aviation, in the 1920s and 1930s, was a small community, and there were relatively few private pilots. From his own flying, Dad met and became friends with lots of the famous flyers of the day. Often, on their way cross-country, they would buzz the house and put down at Regan field, a half-mile from our house. We'd drive over and pick them up, and our guests would stay for dinner and usually spend the night.

In this way, I met Amelia Earhart.

She was a beautiful woman and very nice to me, and I was smitten with her, although I couldn't have been more than nine or ten at the time. She was quite ambitious, but the pilots I heard talking around the house didn't consider Amelia a particularly skilled flyer. In their view, mostly what she had was a burning desire to make a name for herself, and always enough money to buy the newest and best aircraft.

Amelia was rather shy and was known for having an aversion to speaking over the radio. I've often wondered if her dislike for radio communications helped to seal her fate. At her last stop in New Guinea on her globe-trotting flight in 1937, she made a critical decision not to wait around a few days to replace the long-line antenna for her automatic direction finder (ADF) and high-frequency radio, which would have given her eight to ten times the range she had with her shorter radio antenna. That meant that she and her navigator, Fred Noonan—also a friend of Dad's and occasional guest at our home—were severely penalized on the longest over-water leg of the trip when it came to trying to get an ADF fix from tiny Howland Island. Noonan was an experienced navigator and used a procedure common in those days called the "offset technique." You flew the proper distance to your destination, bearing slightly left of course, then did a 90-degree right turn and continued on until you hit your destination. This method didn't require as much precision about latitude. Had you aimed directly for your destination and not found it, you couldn't be sure whether to turn left or right to locate it and could end up circling aimlessly. With the offset method, you anticipated the right-turn correction by purposely flying to the left of your destination. The conjecture among pilots was that Noonan probably went too far to the left, and without being able to pick up a navigational fix from Howland, they ran out of gas before they could find the island. If Amelia had placed more value on the radio and had waited for the long-line antenna, they might have had enough range to get a fix and contact a ship- or shore-based radio operator.

Another of Dad's flying friends who stopped in Shawnee whenever he was passing over and often showed up for dinner was Roscoe Turner, one of the most flamboyant and successful pilots of the scarf-and-goggles era. His flair for self-promotion combined with icy nerves and flying skills made him one of the most successful competitors in the air races of the 1930s; seven times he broke transcontinental speed records. A real showman, Roscoe was partial to leather pants, knee-high riding boots, and long silk scarfs. Once, when he stopped over on his way to an air show, he had a real live lion cub named Gilmore flying with him. Many years later, shortly after my Mercury mission, I attended the Indy 500 race, and on race night Roscoe hosted a lavish party at his estate in Indianapolis. I walked into the living room and the first thing I saw was Gilmore—full-sized and *stuffed*.

It was a real tragedy around our house when famed aviator Wiley Post and humorist Will Rogers died in a plane crash in Alaska in 1935, an event that riveted national attention every bit as much as the John F. Kennedy, Jr., airplane crash. Wiley had been a guest at our house on several occasions, and we all liked him for his unpretentious ole-country-boy humor and endless flying stories. I even had the opportunity to go up for a spin with him; Dad considered Wiley one of the best pilots he'd ever met.

Wiley, who had once been told he would never fly because he had only one eye (from an oil-field accident), was the first pilot to circle the world solo. He also ended up setting a slew of altitude and distance records in his Lockheed Vega, named *Winnie Mae*. In fact, he developed a pressure suit for his pioneering high-altitude flights, becoming the first pilot to wear one, and blazing the way for my generation of pilots and astronauts.

After retiring *Winnie Mae*, Wiley assembled a new aircraft from the fuselage of a secondhand Lockheed Orion and the wings of an experimental Lockheed Explorer. Both planes had previously been involved in accidents. Wiley added a three-bladed propeller and six tanks, which carried 270 gallons of fuel. For the Alaska trip—Will Rogers wanted to interview an old trader in the bush country for his popular syndicated newspaper column—Wiley replaced the plane's wheels with a set of floats from a German-built Fokker trimotor for water landings. Wiley hadn't had much time to get accustomed to his customized aircraft's flight characteristics, and he found that at low air speeds and low power settings his new plane developed an uncontrollable, nose-

down pitching attitude that concerned him, but he was commit-
ted to making the trip. As for Will Rogers, who hailed from
Claremore, about sixty miles north of Shawnee, he felt secure
with his fellow Oklahoman at the controls.

In Alaska, they found themselves in zero-visibility weather,
and Wiley flew for a couple of hours on instruments. They
dropped down through a two-hundred-foot ceiling over a river in
driving snow and ice and landed on the river to ask some Eskimos
which way it was to Point Barrow. The Eskimos pointed the way,
and Wiley took off. As he was pulling into the cloud cover, the
plane stalled, rolled over, and came down at a high rate of speed,
crashing into the river and flipping over. Wiley, pinned underneath the engine, was killed instantly. Will Rogers, pulled from the
wreckage by the Eskimos, was dead too.

Pancho Barnes, proprietor of the famed Happy Bottom
Riding Club at Edwards Air Force Base, was another old friend
of Dad's. Her club, frequented by all the top test pilots, who
were breaking sound-barrier records on a regular basis, had
become one of Dad's favorite hangouts whenever he stopped off
at Edwards, and Pancho had visited our home, where my
mother accepted her with open arms. In fact, Pancho was practically like an aunt to me, and I always looked up to her as a
pilot because she'd done so many great things. (In 1930 she
broke Amelia Earhart's world speed record.)

The daughter of a minister, Pancho had married into
wealth but was married in name only. Each went their own way,
and for Pancho that meant flying and raising hell with other
pilots. In the aviation community Pancho was considered a top
pilot, with lots of experience in various aircraft. Spontaneous,
adventurous, and dedicated to flying and friends, Pancho also
had a big heart. Many down-and-out aviators stopping off at her
club, where she had cottages out back for overnight guests, got
their planes filled up with gas and a hot meal from her, gratis,
when they needed it.

In 1953 the air force condemned Pancho's land to build a new runway. Pancho suggested in no uncertain terms that they build the runway somewhere else. The air force dug in its heels, and Pancho filed suit against the commanding general in federal court in an effort to stay in business. Then working as the JAG officer for the Air Research and Development Command in Baltimore, Maryland, Dad saw the futility of Pancho's legal situation and flew out to California to have a heart-to-heart talk with her. He told her the runway was going to be built and she might as well drop her suit and save herself the grief. He also let her know that the air force had film of scantily clad young women running around the place, partying with pilots. Pancho, a stout woman given to wearing flannel shirts and jeans, had never run a house of prostitution, but things could get dirty in court. She refused to give quarter, and a lengthy legal battle ensued. Although she ended up collecting payment for the value of her land, she lost her place of business, as Dad knew she would. By the time I arrived at Edwards in 1956, the club had been razed, the new runway built, and Pancho was long gone.

Two decades later, at the 1975 rollout of the newly built B–1 bomber, Pancho was expected to attend and I was looking forward to seeing her. She never showed up. We didn't know until later that she'd suffered a fatal heart attack a day earlier and her body was still undiscovered at home. At the time of her death at the age of seventy-four, she had been living on and off with her fourth or fifth husband, who was in his mid-twenties. Friends obtained permission from the air force to fly over the site of the Happy Bottom Riding Club and scatter her ashes.

By the time I was old enough to join the military (seventeen with parental permission) during World War II, there was a surplus of pilots, and no service would guarantee flight training to any new recruits, even those who already knew how to fly. So I enlisted in the Marine Corps, which I'd heard was the best fighting infantry in the world, hoping to do my part as a rifleman. In so doing, I turned my back on the possibility of a football scholarship to Oklahoma A&M. I had played a lot of sports in high school and was graced with natural speed—I was a sprinter (ten-flat in the hundred-yard dash) and quarter-miler (low fifties) on the track team, and my high school football team played for the state championship my senior year with me at halfback.

Since I was still in high school when I enlisted, I was not called to active duty until I graduated in June 1945, and the war ended before I saw combat. I passed some tests and was assigned to the Naval Academy Prep School as an alternate for an Annapolis appointment. The guy who was the primary appointee made the grade, however, so I was reassigned and wound up in Washington, D.C., serving with the Presidential Honor Guard and occasionally having coffee with President Truman, who at night would sneak away from the White House without his Secret Service guards to take a stroll, and often ended up at the marine barracks five miles away. "No formalities, son," he'd say. "Just a cup of coffee." A real down-to-earth guy, he'd tell us stories about his days as a combat artillery officer during World War I, and we chatted back and forth just like a bull session at the American Legion. Then he'd thank us, put on his hat, and go out the back door for his solitary walk home.

After my discharge in 1946, I went to visit my parents in Hawaii, where Dad was then stationed. I ended up staying and enrolling at the University of Hawaii, where I joined the Army ROTC. I majored in civil engineering—the only engineering major offered. Soon I had my own practically new J–3 Cub, signed over to me by a couple who were moving to the mainland and found the cost of shipping the plane prohibitive. (In turn, when I left Hawaii I signed over the plane to another young pilot.) The islands were a great place to fly, and I spent about as much time in the air as I did on the ground.

There were many other distractions, too—the daily temptation to goof off from serious study beckoned from every accessible beach. Every grain of sand, every wave, and every sun ray were siren calls. I did well in humanities but began to suffer in engineering, where the competition was fierce, particularly from Asian students. I knew that just being able to fly wasn't enough, though, and if I ever doubted that, my parents—big believers in education—would set me straight. Whether I stayed in the military or went back into civilian life, they told me, I would need my degree to get a good job. I tried to stick it out, but these were not my most stellar times academically.

It was in Hawaii that I met Trudy, a beautiful blue-eyed blond of Norwegian descent. A licensed pilot, she was quite active in flying and was planning to complete the requirements for an instructor's license so she could teach flying. She also loved the beach and surfing, my other passions, and I became very intrigued with her. After a six-month courtship, we married. We soon had the first of our two daughters, Camala, with Janita to follow a year later. Trudy and I were married so young (both twenty) that the odds were stacked against us. As we grew up together, we sadly grew apart.

Meanwhile, the ROTC led to an army commission, which, in 1949, I was able to transfer to the air force when they put out the first call for flying school candidates since World War II.

When I got into flight training, all my flying paid off. The first military aircraft I flew was a T–6, a high-performance single-engine trainer that was a real handful if you hadn't flown before. The T–6 was unforgiving. It wouldn't just float along and let you be lazy; you had to fly it all the time. It was a snap for me because of my experience in so many different aircraft, and I loved flying it. You had to get past the T–6 before moving on to other types of aircraft, and a lot of cadets washed out during this phase.

I had not mentioned my flying experience to anyone—I preferred to show what I could do rather than talk about it. My instructor, a tough veteran of the famed Eagle Squadron in England, where he flew Spitfires in dogfights against the best of the Luftwaffe, took me up and went through a few routine maneuvers, then told me to "try a few things." I took the stick, and he didn't get it back that day.

"How much flying time you have?" he finally asked.

"Quite a bit."

"It shows. Let's do all the required procedures. Then I'm gonna turn you loose."

We went through a few takeoffs, landings, and stall recoveries. A stall has nothing to do with the engine: You take the plane straight up until it slows to a stop and starts falling from the sky. Recovery involves putting the plane's nose down and building speed up before regaining control and being able to return to level flight. I also demonstrated basic acrobatics such as spins, loops, and rolls.

My instructor soloed me on my third flight. I had to meet all the requirements at various stages of the training, but otherwise I'd go out and practice whatever I wanted.

For the first time in my life I was getting paid to fly.

I knew I'd found my life's calling.

We are all shaped to some degree by our parents. In my case, I wonder how my life would have been different without my father's love of flying. Without him, I might never have discovered my truest element.

Flying has a calming influence on me. People who know me say that I'm a totally different person when I get in the air. I think they mean that I'm clearly happier and easier to be around when I'm flying than I am sometimes on the ground. I do tend to get cranky when I go too long without flying. Luckily, I've been able to spend a sizable portion of my waking hours airborne.

I live for that exhilarating moment when I'm in an airplane rushing down the runway and pull back on the stick and feel lift under its wings. It's a magical feeling to climb toward the heavens, seeing objects and people on the ground grow smaller and more insignificant. You have left that world beneath you. You are inside the sky.

Flying is not just a fair-weather exercise. Some of my most thrilling moments have been flying at night or in heavy overcast, when you might find yourself yawing uncontrollably without realizing it or even flying upside down. What is needed is a steady hand and a clear head—and also, from time to time, some old-fashioned luck.

Not long after my return to space aboard *Gemini,* I had a close call in a twin Bonanza. It was owned by Jim Rathmann, who operated a General Motors automobile dealership near Cape Canaveral and had won the Indy 500 in 1960 with an average speed of 138 miles per hour. Jim had taken the astronaut corps under his wing and helped us get the deals on our Corvettes. (One Mercury astronaut hadn't been interested: John Glenn, who stubbornly kept his Prinz, a kind of mini-VW bug made in Germany, which would have had a hard time beating the meter maid's scooter.)

I was pilot of the Bonanza, and Charles "Pete" Conrad, one of the nine test pilots who had been named astronauts in 1962, was co-pilot. We were something like number eighty-six in line for takeoff at the local airport after the conclusion of the Daytona 500, and while we waited the weather kept getting worse until the visibility and ceiling dropped below minimums for visual flight rules (VFR). Since the aircraft had all the necessary equipment for instrument flight, Pete got busy scribbling out an IFR plan, and filed it by radio with Flight Service by the time the tower notified us we were number one to depart.

After takeoff, the tower said, "Turn right 210 degrees. Climb to thirty-five hundred and intersect Melbourne VOR."

I acknowledged, and turned right.

All of a sudden, we surged upward at a tremendous rate of

climb. I felt it in the pit of my stomach before checking the altimeter and seeing the hand spinning.

Up up up we were being blown by the roaring and howling wind. It was something I'd never experienced before. Then, just as rapidly, we were being pushed downward into a deep dark pit. I couldn't see a thing out the cockpit window. Daytime had suddenly turned to midnight.

Realizing we didn't have much altitude as yet, I did everything I could to slow the descent rate. I had my hands full. Next to me, Pete was hanging on for dear life, and in the backseat, Rathmann, the speed demon, was sure we were about to die.

It turned out Flight Service had vectored us into a tornado.

What saved us was that Beechcraft builds twin Bonanzas like tanks. Any other type of plane could easily have shed its wings in such ferocious winds. Somehow we got through it, although in spots the paint was peeled off the fuselage.

Such adventures are what pilots live to talk about with their flying buddies.

My father knew about that.

A reader of Buck Rogers and Flash Gordon, Dad also knew that one day man would be traveling in space. He didn't know when, mind you, but he *knew*. And he shared his vision with me as we flew side by side. I used to wonder if all this was going to happen in my lifetime or if the world was going to have to wait until something like the twenty-fifth century before people might actually get out into space—not just for the thrill of it but to explore and discover new things and find out what the planets were really like.

A year and a half before I was selected for the space program, Dad was diagnosed with terminal lung cancer. Given six months to live, he said the hell with the doctors and headed for the high country, where he climbed on a horse and went fishing.

Dad was a great fly fisherman. I was his apprentice—he bought me my first fly rod when I was four years old—and became a good fisherman myself, but I never measured up to Dad. He could put a hundred feet of line out upstream and land the fly exactly where he wanted, softly, surely, deliciously. It looked so much like the real thing that in some of those crystal-clear mountain streams of Colorado, a big trout would hit his dry fly before it struck the water.

Three years after Dad's diagnosis, he was still doing OK. He and Mom came out and spent Christmas 1959 with Trudy and the girls and me. I showed him around NASA and introduced him to the guys: Gus, Al, John, Scott, Wally, and Deke. I told him all about the weightless flying training we had begun that very month: going up in a two-seater F–100, and as the pilot put it into zero-gravity maneuvers, sitting in the back, eating, drinking, and testing various motor skills.

Dad was mighty proud of me, and in general, excited about the space program. I think there was also a tad of envy—wishing he was twenty-five years younger, getting ready to go into space.

On their way home from that trip, Dad had some kind of seizure and lapsed into unconsciousness while Mom was driving. She raced him to a hospital, where she was told that the cancer had spread to his brain.

I took emergency leave, and once we got Dad home, he went to bed for good. He could move only one hand and one foot, and his face was contorted by paralysis. He lingered for about a month and was never able to talk again. Sometimes his eyes were open and would follow me around the room. He was in and out of a coma. At times I thought he understood what we were saying, but I couldn't be sure.

I was with him when he died on March 29, 1960. He was fifty-eight years old. He was given a full military funeral and buried in a small cemetery in Colorado that looked like Boot Hill. It was a beautiful setting on the brow of a hill overlooking the valley where our cabin was situated, surrounded by snow-covered mountain peaks.

The air force sent a color guard, and a bugler sounded taps.

Dad never got to see me go into space. As a matter of fact, he didn't live to see any man go into space. He passed away a full year before our first mission.

But somehow I was sure he was watching.

BACK TO SPACE

My return to space began in the jungles of Panama.

I was scheduled to be command pilot of *Gemini 5,* a two-man spacecraft that was heavier, larger, and more advanced than Mercury. What we had accomplished in having a test-flown two-man spacecraft by 1965, having started from scratch seven years earlier, showed how hard a bunch of dedicated and talented people were working.

I would be going into space with U.S. Navy Commander Pete Conrad, one of the second wave of NASA astronauts, dubbed the "Gemini Nine." A go-getter-type guy, Pete was thirty-five (three years younger than me), the shortest of the astronauts, at five feet six inches, and also one of the funniest and most irrepressible with his lively sense of humor and gap-toothed grin. Years later, when Pete made it to the Moon—one mission after Neil Armstrong's historic first step—his comment to the world as he climbed down the lunar module's ladder and pushed away from the bottom rung for the considerable drop to the lunar surface: "That may have been one small step for Neil, but it was a long one for *me.*"

We were to be propelled into orbit by the powerful Titan II rocket, modified for manned space work from a second-generation intercontinental ballistic missile built to deliver hydrogen bombs into the heartland of the Soviet Union in the event of all-out war.

But first came the jungle.

Pete and I were dropped by helicopter into Panama for several days of survival practice—in case we missed our ocean landing spot and ended up in some place like Borneo. The chopper set down in a small clearing atop a mountain ridge and let us out with only what we'd have in our spacecraft. They also dropped the wooden-and-cloth frame of a Gemini spacecraft, as well as a deployed main chute so we'd have the nylon canopy and shroud lines to use in our survival efforts. We were wearing the outer shell of our space suit, although we took it off mighty quickly in the heat and humidity, cutting down our long underwear into more suitable jungle attire.

We were to be picked up at this same spot in five days. If something went wrong, such as an illness or injury, we had our emergency survival radio and could call out for an immediate pickup.

For several days previously, Pete and I had attended classes at the U.S. Air Force's jungle survival training center in Panama, learning from experts about what to expect in the jungle. We were shown what flora and game to eat and what to avoid. One afternoon, as we listened to our instructor, a huge boa constrictor was released into the room and slithered quietly under the desks. We knew it was there only when it lifted its big head three feet off the floor and looked at us. I had always liked snakes, and I got along so well with our classroom visitor that, before leaving Panama, I was presented with one of her babies (she'd recently had a litter), which I carried home in my briefcase.

We took into the jungle with us the regular Gemini survival gear, which included several quarts of water, food rations (dehydrated meals), a machete, a fire-starting kit with magnesium spark ignitor, and cotton balls for getting a fire going. We each carried a survival knife, designed especially for the space program by Bo Randall, a leading knife expert. These steel-forged knives were very strong and could be used for many chores.

We found it easy to survive in the jungle as long as you don't mind eating anything you can catch. We were able to lure flesh-eating piranha into some little streams (OK, I used Pete as bait) and catch them before they got us. We used safety pins for hooks, line from our survival kit, and insects as bait. The piranha grilled up real nice on an open fire. We also developed a taste for lizards and snakes, boiled or grilled. Basically, we ate whatever was slower than we were.

It rained practically nonstop, giving true meaning to the term *rain forest*. At night it got miserably cold. To build a shelter, we cut down small palm trees and used the trunks for the framework, then wove together giant palm fronds for the roof, under which we slung our nylon hammocks. That first cold and wet night, I pulled out of my gear an airplane-size bottle of Grand Marnier. We sat in our hammocks passing the miniature bottle back and forth, taking tiny sips so it would last as long as possible.

One morning Pete and I were strolling along the top of a narrow ridge when suddenly, appearing from the jungle canopy like apparitions, came five bare-chested Choko Indians wearing only loincloths, armed with blowguns. They were pygmy in stature—about up to our chests—but well proportioned and clearly no one to mess with.

We stood face-to-face.

Pete and I showed the biggest smiles we could come up with.

The Indians spoke Spanish—as did I—and turned out to be very friendly. When I explained why we were there, and what it was in preparation for, they looked up at the sky and back at us with grave concern. I believe they were concerned about what we'd been eating or perhaps smoking out there in the jungle.

Each of them had a leather thong around his waist from which hung a machete so long that it practically dragged on the ground. On the other side they had a couple of small leather pouches in which, we learned, they carried darts tipped with curare, an herbal nerve poison that could kill large game within seconds, and smoke-dried meats to sustain them on their hunting expeditions.

Our new friends made a point of hanging around for a few days to show us the ways of the jungle. It amazed me how they could down a game bird in midflight with one strong exhalation into a blowgun. I asked cautiously if they ever used the poison-tipped darts in battle. They said it was very rare for them to fight other tribes, which, considering their accuracy with the blowguns, was a good thing.

They also showed us how to set snares for small game, such as the paca, an oversized rat with a beaver-shaped head that weighed about thirty pounds. We'd been avoiding these rodents, thinking they were not something we wished to tangle with. But the meat turned out to be tasty, not unlike venison. Our menu improved with the help of our jungle tutors, expanding into a bountiful feast of wild game, fruits, and salads. Pete and I must have been the first astronauts to gain weight in jungle survival.

It turned out that the leader of our small pack of hunters was the chief of his tribe. We were given a royal reception in their village, where they lived in huts made of grass and palm fronds, built much better than our own but still temporary. A nomadic people, they lived in one place for a month or two, then moved on, like countless generations of their ancestors before them. Before we left, we received a nice send-off from the village, which included gifts, singing, and ceremonial dancing.

For years afterward, and as a result of our encounter with them, this tribe continued to help survival teams from the air force survival training center in Panama. In recognition for the helping hand they showed so many of our pilots and air crews, the chief of the tribe—the same chief we had come across on the ridge top that day—was summoned to Washington, D.C., in the early 1980s and given a medal by President Bush.

Beginning with Gemini, NASA administrator Jim Webb decided he didn't want U.S. spacecraft to have names. Now we all liked Webb and admired his abilities, but we *did* find ourselves arguing with him a lot. His decision to "depersonalize" the space program was our biggest argument ever. He wanted the flights referred to by numbers—"I want the flights all very machine oriented," Webb explained. It was a serious miscalculation on his part because the personalized stuff—not the science or electronic advances—was what the public identified with and enjoyed the most.

We kept trying to get Webb to change his mind. Pete came up with the notion of naming our spacecraft *Lady Bird* in honor of President Johnson's wife. We figured there was a good chance it would be approved, given that we wanted to name it after the First Lady. But NASA turned us down.

Several months before the mission, I mentioned to Pete that I'd never been in a military organization that didn't have its own patch. Pete hadn't either. We decided right then and there that we were at least going to have a patch for our flight.

Pete's father-in-law had whittled a model of a Conestoga wagon, the preferred mode of transportation for pioneers of an earlier era. We thought a covered wagon might be a good way to symbolize the pioneering nature of our flight. Since our mission was designed to last eight days, the longest ever attempted by the United States or the Soviet Union, we came up with the slogan "8 Days or Bust," which we overlaid on a Conestoga wagon. We gave the design to a local patch company, and they produced hundreds of them. Pete and I had ours sewn on the right breast of our space suits.

Two days before launch, Jim Webb, in from Washington,

beckoned us to Houston for what was to become a prelaunch tradition: dinner and a social evening for the prime crew at the home of Bob Gilruth,who had been appointed head of the new Manned Space Center.

Pete and I climbed into a T–38 and flew to Houston.

The Mission Control Center had moved from the Cape to Houston, and ours would be the second manned flight to be controlled from the new location. The move had been pure pork barrel, offered up by the powerful Lyndon Johnson to his home state while he was still Senate Majority Leader. Several billion dollars were spent putting the Manned Space Center in Houston. Hundreds of millions more in added costs were associated with having Launch Control in Florida (early in the U.S. space program it was decided the rockets had to be launched over water to avoid populated areas) and Mission Control, which took over control of flights immediately after launch, located a thousand miles away in Texas. Simulators and other expensive systems had to be duplicated, not to mention astronauts and technicians constantly burning up the airways as they ran back and forth between the Cape and Houston. In fairness, other locations for the new space center had been considered, and Houston had met all the criteria. Such requirements included being on the coast and so able to receive equipment and other heavy materials by ship, and being near a major university able to provide a pool of qualified scientific consultants.

That night during dinner, I decided we had to tell Webb about our "8 Days or Bust" patch because it wasn't fair for him to find out by surprise or through the media.

"Jim, you've taken our spacecraft names away from us, and as you know, none of us particularly like it," I said. "Pete and I want to personalize our flight, and we've designed a really neat mission patch."

Webb about went into hysterics. The patch was in direct violation of his efforts to depersonalize the space program. The argument got so heated that at one point Bob Gilruth and I had to pull Webb and Pete apart—the overall head of NASA and one of his astronauts were stopped just short of fisticuffs.

When Webb cooled down, I explained how Pete and I had never been in any military organization that didn't have a patch. "It's not just for the guys flying," I went on, "but for the hundreds of people working on the launch equipment and operating the worldwide tracking range and all the other things that go into a successful mission. Wearing that patch tells the world that they worked *Gemini 5*."

Webb asked me if I had the patch with me.

Unfortunately, we hadn't thought to bring one.

He asked that one be flown to Washington the next day. "I'll look at it and make a decision," he said.

"Fair enough, Jim."

The next day, after reviewing the patch, Webb called me at the Cape. "All right, I'll approve this patch on one condition."

"What's that?"

"That you cover the '8 Days or Bust' until you make the eight days. If you don't make eight days, I don't want the press having a field day about the mission being a bust."

So we had little pieces of canvas lightly sewn over the offending slogan.

In his official memo authorizing the patch, Webb directed that all future space crews could have their own mission patch, henceforth to be "referred to by the generic name of the Cooper patch"—a tradition that still lives today.

Come launch morning, Pete and I found ourselves sitting in the spacecraft—*Gemini* was about the size of a VW beetle lacking a backseat—on a late countdown hold waiting for some last-minute problem to be solved. This was our most vulnerable time, the last twenty minutes of the countdown, because the final firing sequence had begun to take things over automatically; all systems had been activated and were ready to go.

All of a sudden a big thunderstorm moved in. Although by then we had weather radar at the Cape, weather forecasting in southern Florida proved to be a tough job. Weather fronts and thunderstorms develop very quickly in that latitude and had a way of catching everyone—even the meteorologists—off guard.

Before it could be decided whether to continue the wait or scrub the mission and shut everything down, lightning hit a main power cable, knocking out power to our pad and other systems. Without power, Mission Control was unable to deactivate the rocket, and there we sat—stuck in the middle of the firing sequence and on top of a hot booster—with a lightning and thunder show all around us.

Now there was no choice in the matter: the circumstances dictated scrubbing the mission. With a loss of power so late in the countdown, there were too many systems that needed to be reset and recycled. We would not be going into space this day. The only question now was when and how we would get out of the spacecraft. Without power, Launch Control couldn't bring the gantry, which rolled under power along railroad tracks, back up to the rocket to get us out via its twelve-story elevator.

So we sat there on the hot booster for an hour, waiting for power to be restored, knowing that if there was a premature launch or explosion, we could be in a pickle.

In Mercury, the spacecraft had a sixteen-foot rocket-propelled escape tower attached to the top of the spacecraft, which was designed to fire in an emergency situation and pull the spacecraft free of the booster. But we had no such escape tower for Gemini. The reason: at one point, we had hoped to "fly" *Gemini* to a landing with a collapsible fabric wing that would be deployed once we were back in the Earth's atmosphere. Retractable landing gear had even been installed, and our notion had been to glide home like a sailplane—as the space shuttle does today.

We tried some test flights, and the wing worked beautifully. However, it had to be folded up and stored into such a small

canister and at such high pressure that we kept having problems with folds and rips in the fabric. The best technicians
NASA had on the payroll tried to solve the problems but
couldn't. Finally, we had to give up, and practically with tears in
our eyes we had the landing gear removed and shelved the wing.

The spacecraft had been designed with two ejection seats,
meant to provide an airplane-type escape in the event of trouble
with the landing. The seats had remained because it was too late
to reconfigure the spacecraft with an escape tower. *Theoretically*, if
everything went just right, the ejection seats would do the job and
pull us away from a launch-pad emergency, but no one had ever
tried it, and we sure didn't want to be the first.

It was finally decided to roll out the emergency cherrypicker, a self-powered vehicle that looked like one of those utility
company trucks but had a ladder that went much farther into
the sky. It had been designed to evacuate astronauts in an emergency but as yet had never been used under real conditions.

Not long before Al Shepard's first flight, the flight director
had approached me and said, a bit chagrined, that there were no
arrangements in place for getting the astronaut out of the capsule in case of trouble on the launch pad during the countdown.
Up to then all the focus had been on what happened during
flight. I hurriedly went to work setting up emergency procedures—including fire-fighting crews and techniques for handling the volatile fuels and the various pyrotechnics like the
escape-tower rockets and retro rockets. If the spacecraft toppled
to the ground in a fire, for instance, an armed personnel carrier
would roll out, manned by a special team that included technicians who knew where all the seams and bolts were. Wearing
fireproof suits, they would extract the astronaut. For Al's suborbital flight aboard the Redstone booster, I had stayed in the
blockhouse, standing by to help put into effect any possible pad
rescue operation as well as handling communications to the
spacecraft.

Starting with Atlas and the third Mercury mission, we had a tower built just twenty-five feet away from the gantry with a special drawbridge attached, which could be lowered in thirty seconds so that the end of it sat just outside the spacecraft's hatch. If an emergency arose, the astronaut could blow off the hatch, scoot across the gangplank, and take an express elevator on the tower that would get him to the ground in thirty seconds. By then we could have the personnel carrier on the spot to pick him up and get him clear of the area. With the tower so close by, we hadn't felt we'd need the cherrypicker anymore. Still, we decided to station it behind the blockhouse just in case—and it was a good thing we did. Without electricity for the elevator, Plan A for getting out of the spacecraft wouldn't work.

When the cherrypicker got close enough to the venting rocket, two technicians climbed to the top. Using a speed wrench, they unbolted the hatch—freeing us in five minutes. We weren't so much relieved as disappointed at not getting off as planned.

That night Pete and I had a couple of beers in Hangar S, trying to uncork the built-up and unused adrenaline. As we talked about our prospects for launching the next morning and tried to relax, technicians worked through the night shutting down the rocket, topping it off with fuel, and recycling and resetting all its electrical systems.

Even though we were designated *Gemini 5*, we were the third manned Gemini mission. After two unmanned Gemini launches, the first manned mission had gone to my old buddy, Gus Grissom, who became the first person to fly in space twice, and one of the newer astronauts, John Young, a navy pilot.

Still smarting from the sinking of *Liberty Bell 7*, Gus had wanted to name his Gemini spacecraft *Molly Brown*, after the popular Broadway musical comedy of the day *The Unsinkable Molly Brown*. His second choice: *Titanic*. Then came Jim Webb's edict, and NASA mandated that the mission be referred to simply as Gemini 3.

I was serving as CapCom on launch day, March 23, 1965.

At liftoff, Gus reported, "The clock has started."

"Roger," I replied. "You're on your way, *Molly Brown*."

From then on, to the press and the rest of the world, the Gemini 3 spacecraft was known as *Molly Brown*, much to the displeasure of the NASA bosses.

Gemini 4, with Ed White and James McDivitt, went up on June 3, 1965, and accomplished another historic first: the first U.S. walk in space. Ed joked from outside the spacecraft as he floated in space on a tether: "I'm not coming back in."

In all, there would be twelve Gemini flights, with three Mercury astronauts (Wally Schirra in addition to Gus and me) thrown into the mix of newer astronauts to fill out the flight crews. Of the original Mercury Seven, Deke Slayton was still grounded; Al Shepard's flying career was on hold because of an inner-ear problem that affected his balance; John Glenn had already quit the space program to go into business and politics; and Scott Carpenter was busy exploring oceans around the world.

It was no accident that Pete and I were scheduled to stay in orbit for eight days: that was the longest time it would take to fly to the Moon, explore its surface (for several hours), and return. As part of our research and development for the upcoming Apollo lunar missions, we would be testing and proving not only the onboard equipment and flight maneuvers—such as rendezvous and docking—necessary for missions to the Moon but also man's capability to stay in space for several days and withstand the effects of extended weightlessness.

We launched the following morning, August 21, 1965, at 8:59 A.M., from pad 19—one of two Gemini pads at the Cape. Liftoff was smooth, and our trajectory was almost perfect. I found the ride atop the Titan II considerably less bumpy and more solid than my ride on the Atlas two years earlier. In fact, compared with the thin-skinned Atlas, the Titan, a solid, thick-walled booster, was like cruising down the road in a Cadillac.

The next thing I noticed was how much quieter Gemini was than Mercury, which had a high noise level because the inverters, motors, and all the other things were running so close by. For Gemini, we had moved all of these systems out into a separate "adapter section" of the spacecraft. While there was actually less space per person than we had in Mercury, it was more strategically arranged in Gemini so that it felt like more usable room. On Gemini, we also had storage lockers right in back of us; we could squirm around and maneuver things in and out of the lockers, which made for a less cluttered cabin.

Ours was the first spacecraft to go into space with a fuel cell: an on-site electrochemical generator that produced its own energy. Previous spacecraft had relied on batteries, which would be too cumbersome and heavy given the amount of electronics the more advanced spacecraft were now carrying. In *Gemini 5*, for example, we were taking into space the first onboard radar, and first computer, both of which drew substantial electric power.

Proving that we could fly with a fuel cell was paramount. Unlike batteries, which consume themselves, fuel cells are almost endlessly rechargeable as long as there is a fuel source to provide hydrogen. A fuel cell is operated by the chemical reaction of hydrogen and oxygen and requires that high pressure be maintained in cryogenic (low-temperature) storage tanks so that sufficient amounts of fuel can be stored to provide high electrical output. Fuel cells had been around since the mid-1800s, but thanks to the space program the technology was finally being seen as a primary source of electrical power. The ramifications were awesome, not only for space travel but for life on Earth, with fossil fuels rapidly being depleted. (Today they have gained widespread use in both space and industrial uses.)

On the third orbit, our fuel cell nearly cost us the mission. The oxygen pressure going to the fuel cell dropped from eight hundred to seventy pounds per square inch, and we had no idea why. According to our set emergency procedures, if the pressure went that low on the fuel cell I was required to start shutting everything off to conserve energy.

At the time we were right in the middle of one of our seventeen planned experiments and using a lot of power. We'd released a rendezvous pod, had it on radar, and were just getting ready to intercept it—an experiment designed to provide crucial information about never-before-attempted space rendezvous.

When the trouble occurred, we were out of communication with Mission Control, *of course*. Based on what I was seeing on the panel, I had no choice but to abandon the rendezvous pod, forget the experiment, and start powering down.

When we came into communication again, we were in a drifting mode.

I reported the problem and gave them the numbers.

For some time, it was touch and go with Mission Control. They were close to ordering an emergency reentry, thereby ending our ambitious mission after only three orbits. The loss to the program would have been incalculable, a cost of millions of dollars but also a major setback to our goal of reaching the Moon by the end of the decade.

The pressure that had been put on NASA as a result of President Kennedy's announced national goal of landing a man on the Moon and bringing him safely back to Earth before the end of the decade had not diminished. At the time when President Kennedy had first captured the nation's imagination—"We don't choose to go to the Moon because it is easy . . . but because it is hard"—NASA's most optimistic time frame for the first manned lunar landing was 1972. We had since made great strides toward JFK's goal—a political deadline, not a scientific or aeronautical one—but testing was still being hurried, flights consolidated, and corners cut.

We had our share of such pressures during preparations for

Gemini 5. During the last few days that our spacecraft was at McDonnell Aircraft in St. Louis before being shipped to the Cape, we went to St. Louis for some final finishing-up altitude chamber tests. Pete and I had completed the regular tests on Friday but wanted to stay over on the weekend to do some tests with the fuel cell at altitude: running it at drastically reduced pressure to see how it performed. Since this was going to be the first fuel cell in space, we believed these extra few tests were justified.

NASA balked at our plan. The schedule was to ship the spacecraft to the Cape on Sunday. If we waited and did the tests on Monday, we would lose two or three days. And if we did the tests on Saturday, NASA would have to pay double and triple time to the personnel who ran the altitude chamber. When I appealed to Bob Gilruth, he said, "We don't have the money in the budget."

Late that Friday afternoon, Pete and I had gone to see Jim McDonnell and told him about the final tests we wanted to conduct. McDonnell, then in his seventies, was a big supporter of the space program and a dedicated American. He had founded McDonnell Aircraft twenty-five years earlier in a small office upstairs in an old hangar and when World War II broke out started building military aircraft that were to become among the best in the world. I had flown several of his later planes, including the F–101 Voodoo and F–4 Phantom, and found them to be spectacular and well-built aircraft. We hoped he would find a way for NASA to let us do the tests.

He immediately agreed that the tests were important and should be done. "I'll pay for them," McDonnell said in his decisive way. "Let's do it."

Whenever I hear people criticizing aerospace contractors for being greedy, I think of Jim McDonnell, who came to our assistance innumerable times throughout the space program. His unselfish, patriotic, can-do attitude was shared by the vast majority of outside contractors I worked with—including North

American and Rockwell. They were as dedicated to the mission as anyone and acted like partners in this country's space program rather than simply on-hire contractors looking for the next buck.

Now our Gemini mission was saved only by having done those tests and preparing for the worst-case scenario—a replay of the Mercury mission when I had practiced for losing all my electrical power and *did*. I reminded Houston that we had worked with the fuel cell at very low pressure in the altitude chamber and found ways to deal with it.

The guys in Houston quickly located the data on the tests we'd run that weekend in St. Louis and were relieved to see our results with low fuel cell pressure.

Still, the decision whether to bring us back early was almost a flip of the coin.

The fact that the cryogenic tank pressure leveled off at the seventy-pound figure for several hours while the spacecraft was in a powered-down state convinced Flight Director Chris Kraft to let the mission continue for at least one full day while continuing to closely monitor the situation. From then on, a "go" decision was reached each morning.

We soon isolated the problem: the heater that was supposed to be coming on automatically to keep the supercold liquid oxygen for the fuel cell from freezing wasn't working properly. The oxygen wasn't heating up enough to turn into gas, a vital part of the chemical reaction necessary for the fuel cell to produce energy. Once we figured this out, we were able to bring up power slowly for periods of time and begin to warm the partially frozen oxygen in the gaseous section. As that happened, the usable pressure in the tank began to rise.

We took it slowly as the flow of oxygen into the fuel cell gradually increased and the pressure rose. We powered the spacecraft back up over several days—bringing the various systems back on line one at a time. Eventually our control panel was back to looking like a fully lit Christmas tree, and we had power to burn. Proving that we could fly the fuel cell, even under some unusual circumstances, and do so better than anyone had hoped was probably our most important feat.

We had "engineered" ourselves back into a full mission, although we weren't able to go back and do the experiments we had missed on our full flight schedule.

Once we received the "go" for at least a day, we removed our helmets and gloves and stowed them in the foot well, where they would remain until just before retro-fire. Vast improvements had been made over the Mercury spacecraft's cranky temperature control system; we found it very comfortable inside the Gemini cabin. We also donned lightweight headsets with a little boom microphone attached.

Space has a brilliance to it that doesn't exist on the clearest days I've seen high in the Rockies. Photographs can't do it justice. Some of the guys, including Pete, hadn't believed me when I described the details I had seen of Earth on my Mercury mission, including ships' wakes (what I thought was an aircraft carrier in the Atlantic turned out to be a tugboat and barge), highways, and railroads. Pete, in awe of the wondrous sights, quickly became a believer. I was awed too, even though I'd been here before. To see the full majesty of Earth and her continents in such breathtaking views—knowing that we are only one of countless other planets throughout the galaxies—was a privilege few human beings had experienced.

On our first pass over Communist China, we heard a pinging in our ears that turned out to be high-powered radar tracking us from the ground. Then over the radio came a sweet feminine voice that could have been Tokyo Rose, only in this case I guess it was Peking Peggy: "Good evening, *Gemini 5*. For your pleasure we will play some music." They entertained us with the most beautiful opera music, which gave us an especially good feeling since China had officially complained about our flying over their borders. When it was announced that we would be using cameras with telephoto lenses in space, they had publicly accused us of being "spies in the sky."

To replace the rendezvous-pod experiment, Mission Control directed us on our third day to rendezvous with a theoretical target in a different orbit from ours. Under the rules of the exercise, we were restricted to making four maneuvers during two revolutions. We fired the orbital attitude-maneuver thrusters and changed our orbit—the first U.S. space mission ever to do so—by about fifty miles, making it ellipse-shaped rather than circular, and ended up within one-tenth of a mile of our phantom target.

In another important test of the onboard radar capability during the mission, our radar locked on to a radar transponder at the Cape, and we made some measurements that proved pretty accurate. When the equipment in the spacecraft read 167 miles (to target), the radar at the Cape was reading 170 miles. There was jubilation in Mission Control that we were able to track a ground target at all; some technicians hadn't thought the external radar system would survive the rigors of launch. We had proved that radar ranging in space was possible—a vital element in future space rendezvous and dockings.

Pete and I grabbed another record, this one more dubious: the first defecation in space by U.S. astronauts. There had been "no contingency" for such an event in Mercury—they put you on a low-residue diet for two weeks and hoped for the best. Although I can't remember which one of us went first (either that or I'm not telling), it was always an ordeal—lasting about an hour by the time you got your suit off, long undies down, did the deed into a plastic bag, washed up, and got dressed again. Once, Mission Control radioed and asked for me. Pete properly and very politely responded that I was "indisposed."

We were told by astronomers to expect front-row seats for a regular meteorite shower that occurs in the latter part of every August. It would be the first one to be observed by man from

space. The first night of the showers was a sight to behold—
thousands of meteorites passing under our spacecraft as they
entered the Earth's atmosphere and burned up like falling stars.

We knew there was a chance that a meteorite might strike
our spacecraft, but there was nothing we could do to prevent it
and only hoped that if it happened it would be a small one. We
carried a patch kit with rubber plugs to repair any tiny puncture
holes (*tiny* was the operative word) to try to keep from losing
our cabin pressure. But we were not prepared for what it
sounded like when one actually did hit.

A hard metallic *BANG!*

Pete and I both jumped.

It sounded like a major-league fastball hurled against the
side of our spacecraft, but we knew that it was no bigger than a
grain of sand. If the meteorite had been anywhere near the size
of a baseball, it would have gone right through the side of the
spacecraft—ending, in a nanosecond, our mission and our lives.

Over the course of the next couple of days, we were struck
four or five times. When the spacecraft was dismantled upon its
return to the Cape—every returning spacecraft was taken apart
piece by piece as part of a total engineering report to assess how
it handled the stresses of flight—impressions were found on the
outside wall, as if someone had driven home an ice pick with a
hammer. The meteorites had actually reshaped the outer titanium wall of the spacecraft, pushing in the toughest metal
known to man as much as a quarter-inch. (Titanium takes more
heat with less damage than any metal on Earth.) It seemed
unbelievable that such a small particle had so much energy and
caused such a sound, but these cosmic fastballs were a bit faster
than any Hall of Fame pitcher's—a speed gun would have
clocked them in the range of thirty thousand miles per hour.

At eighty-five hours into the mission, including the some
thirty-three hours I had spent in space in *Faith 7*, I passed the
individual space endurance record of one hundred nineteen

hours and six minutes set by the Russian cosmonaut Valery F. Bykovsky aboard his *Vostok V* spacecraft two years earlier. No one made any fuss about it, least of all me, but I think we all felt a lot of pride. The tide had turned in the space race; we all knew it. The United States was no longer lagging in second place.

When it came to our new onboard computer, we were very careful throughout the flight whenever Houston sent us an upload. Before we would accept it on the computer, we'd have them send us a test to run to make sure the information about our orbits and location and onboard systems was accurate. Only if it checked out would we would let it load into the computer. We could then access the information—just as people do today on their personal computers—on a two-by-six-inch monitor. Despite our caution, we got in trouble following the computer on reentry.

Ours was the first reentry to be guided exclusively by a computerized instrument landing system.

We received from Houston a late retro-fire sequence upload. Pete and I hurried to check it out. It looked good, so we let them upload it to the computer.

In Gemini there was no strict division of labor. One reason we were now flying with two astronauts was because there was so much work to do and so many experiments to carry out that one man would have found the workload overwhelming. Too, we were building toward a three-man crew, which would be needed for Apollo and the lunar missions. I was the command pilot and Pete was the pilot, but we had swapped pilot chores back and forth during the mission because it was important to give Pete some experience flying in space. Beyond that, the computer was on Pete's side, so he had been in charge of keeping the systems going, interfacing with uploads and downloads, and doing the rest of the computer work. The navigational needles and other flight controls were in front of me, so I served primarily as head pilot and would fly reentry.

As we started our reentry, I realized we were coming in a lot steeper than we should be. Pete recognized it too, from all our time spent in the simulator. We knew this wouldn't place us in any danger, but it would mean we'd end up way short of our landing spot.

We had made some big improvements in reentry procedures. Mercury simply dropped like a stone until it reached parachute deployment altitude. The pilot's only control came from firing his retro rockets at the precise second to begin reentry. In Gemini, the center of gravity was shifted so that the heat shield was just off center as it met the atmosphere. The result was that the spacecraft had just enough lift to permit the pilot to lengthen his glide path up to three hundred and fifty miles beyond the planned landing site or shorten it by some three hundred miles. Gemini could also be veered to either side of the reentry path by some fifty miles. If the spacecraft was heading on target for splashdown, the pilot just rolled it—about two revolutions per minute—to counteract the natural lift. These features combined to account for some accurate splashdowns during Gemini.

I ignored what the directional needles were calling for— they were following the computer's reentry program—and took control. We were in a communications blackout at the time (of course), but even had we been in touch with Mission Control I'd have done the same thing. In any situation requiring instant response, the pilot flying the machine made the call rather than someone on the ground.

I went with full downrange lift and flattened out our path as best I could. I recovered much of the lost ground, but we still landed a hundred miles short. (Had I not taken over, we'd have been about two hundrd and fifty miles short.) It took the helicopter recovery crew forty-five minutes to get to us.

The recovery team had been trimmed down from previous missions, and since my last return from space, recovery operations had switched to the Atlantic. Our landing target was only four hundred miles off the Cape, making it more convenient to bring the astronauts and spacecraft back home. NASA and the navy had by then gained enough experience to know what was needed and what was unnecessary. The fleet that had been assembled for us was still sizable, and would be the model for future space missions: in all, twenty-eight ships, 135 aircraft, and ten thousand people were involved. The main unit of the recovery fleet was stationed at our prime landing spot in the Atlantic, but other elements were stationed at several locations around the globe in case we needed to end our mission early.

After swimmers went into the water and attached a flotation collar to the spacecraft, they opened the hatch. Pete and I climbed out and were hoisted aboard the helicopter for the forty-minute ride to the deck of the carrier USS *Lake Champlain*.

We then learned what had happened: The genius mathematicians and astronomers who went through all these computer gyrations to give us formulas for our reentry had calculated that the Earth rotates 360 degrees per day, when in fact it rotates 359.999 degrees. When you multiply that times 120 revolutions in eight days, it gets to be a significant figure. So we had loaded into the computer the wrong reentry calculation. It was an early lesson for us all that "computer problems" often begin as human errors.

We had completed 120 revolutions of Earth—a total of 3,312,993 space miles in an elapsed time of 190 hours and 56 minutes. We were 104 minutes short of eight days, but uncovered the "8 Days or Bust" slogan on our patches anyway.

Onboard the ship, I found myself more wobbly than I had been after my Mercury flight, but other than that I felt fine. The eight days of weightlessness had presented no problems; nothing a hot shower, a good meal, and some sleep wouldn't cure.

Gemini 5 carried twenty different types of cameras and several hundred rolls of different types of film, which we experimented with in various lighting situations. Our experiments in surface photography were to identify the problems associated with man's ability to acquire, track, and photograph preselected terrestrial objects from space. We returned with hundreds of great photographs of Earth from space.

One special mounted camera we carried had a huge tele-photo lens. I had been in the military long enough to know what the possible uses of such a telephoto lens from space might be. During an initial flight planning meeting, I raised the issue of whether this was going to be a classified or unclassified project.

"I don't want anyone changing their minds midstream," I said. "If we're going to classify it, let's classify it now."

Nobody seemed to want to classify it, so we went unclassified.

We were asked to shoot three specific targets from our space-craft's window because the photo experts wanted to be able to measure the resolution of the pictures.

That's exactly what we did:

Over Cuba, we took pictures of an airfield.

Over the Pacific Ocean, we took pictures of ships at sea.

Over a big U.S. city, we took pictures of cars in parking lots.

Beyond that, we were encouraged to shoot away at other airfields, cities, and anything else we wanted along the way. That big lens was amazing, and I had a lot of fun with it from our 180-mile-high perch.

After splashdown and while Pete and I were still aboard the recovery vessel, the exposed film from that mounted camera was rushed to a darkroom and developed. I was shown a few pictures—including some unbelievable closeups of car license plates—before someone walked into the wardroom and informed me that all the photos and negatives from that camera were being confiscated and the experiment classified.

I was livid, but there was nothing I could do.

However, when Pete and I went to Washington a couple of

weeks later to receive medals for our mission, I took the opportunity to tell the president of the United States how I felt about
the whole deal.

The loss of John Kennedy had been an incalculable one to
the space program. While Lyndon Johnson assured everyone
that he was equally supportive, we knew he didn't have the total
commitment that JFK had. NASA was concerned, in ways it had
never been during the Kennedy administration, about having its
budget reduced.

I explained to President Johnson, as we sat across from
each other in the Oval Office, that the big-lens photographic
experiment was supposed to have been unclassified. Yet my film
had been taken and I wasn't allowed to see the pictures.

"Son," the president said somberly, "*I* ordered it classified."

The commander-in-chief had spoken, and there was nothing else to say.

Many years later, at a 1997 NASA reunion at Cape Canaveral,
a gray-haired man came up to me and asked if I remembered him.
His face was vaguely familiar as someone who had something to
do with Gemini, but I admitted I was struggling.

"I confiscated your film on *Gemini 5*."

"Now I remember," I said.

"Boy, you were really pissed about that."

I agreed I had been.

"Did anyone ever tell you why the film was confiscated?"
he asked.

"No, I still have no idea. The president said it was classified, and I didn't question that."

The man looked around; we were alone and out of earshot
of anyone. "I'll tell you now because I've heard talk they may
declassify a portion of the film anyway." He paused, but without
losing my rapt attention. "You had the most magnificent pictures of Area 51."

Area 51 was where top-secret black-budget research, development, and flight testing, perhaps using reverse technology from captured extraterrestrial vehicles, was rumored to be taking place despite official denial that it even existed.

When I had worked on the supersecret U–2 program at the secretive North Base of Edwards in 1957, I had picked up hints here and there that it was going to be necessary for the air force to have another area—even more secret—that nobody knew about. I heard it was going to be more remote and easier to safeguard from someone stumbling into it. Entry would be by name and recognition only.

That's probably around the time they started building Area 51 in Nevada's high desert. To this day I have no idea what they were—or are—doing out there because I never talked to anyone who admitted to working at Area 51. Of course, the same could have been said about the highly classified U–2 program when I was with it at Edwards. Nobody talked about the U–2 either, and it remained one of the United States' most closely guarded Cold War secrets until one was shot down on a photo-reconnaissance mission over the Soviet Union.

As for Area 51, I *hope* the air force is conducting experimental flights with highly unusual aircraft—even saucers with revolutionary propulsion systems. I would hope we are getting that kind of value for our money and effort, and that kind of help from advanced civilizations. And the first person to come home with pictures of the mysterious Area 51?

An astronaut from space.

COSMONAUTS AND DEEPEST AFRICA

In the heat of the space race with the Soviet Union, we called a truce.

The astronauts had always wanted to meet the cosmonauts—and vice versa, we were told. We knew our State Department and the Soviet KGB had quietly conspired a few times to try to bring us together but never found a way to do so.

It was amusing, I thought, that the governments of the world's two most powerful nations, while pointing nuclear-tipped missiles at each other and ready to blow one another up, were favorably disposed to getting the astronauts and cosmonauts together. I guess they figured space travelers were something like kindred spirits.

The opportunity presented itself after *Gemini 5,* when Pete Conrad and I were sent to Greece and Africa by President Johnson on a goodwill trip. LBJ even gave us his plane, *Air Force One.* We even got to fly it once in a while.

We left on the day after our White House ceremony and motorcade through the capital. Pete and I had been introduced to a joint session of Congress where we spoke briefly about our mission. I was told I was the first active-duty military man to speak twice to joint sessions—quite an irony considering that NASA had once been so concerned about my public speaking abilities.

We stopped at Athens, Greece, where we had been invited to give talks about our space flight to the 16th International Astronautical Congress, a gathering of space experts from around the world. Before my presentation, Julian Scheer came over. He was the NASA public affairs guy I'd had a scene with when he wanted me to read his prepared speech to Congress after my Mercury flight. By then we had gotten to be good friends.

"Gordo, there's two cosmonauts here. They were the first to go EVA."

"Is that so?"

I knew that the commander of the mission had been Pavel Belyaev—like me a veteran of two space flights. His crewmate, Aleksey Leonov, had been the first, and up to then only, man to go EVA, or open the hatch of the spacecraft and step outside for extra-vehicular activity.

"When you're through talking, just head down the center aisle," Scheer said. "They're sitting in the front row on the right. Walk up and shake hands."

It was a spur-of-the-moment opportunity not to pass up.

When I approached the Russians after my speech, a surprised Belyaev stood, along with Leonov. The three of us shook hands and smiled at one another.

It brought the house down.

That evening, the Greek royals, King Constantine II and Queen Mother Frederica, were hosting an exclusive cocktail party for Pete and me. After the historic handshakes that afternoon, the queen decided at the last minute to invite the cosmonauts to join us.

As Pete and I stood with our Soviet counterparts, one news

photographer was so anxious to get a picture that he fell over a balcony partition and landed with a thud on top of the queen, who was unhurt but not amused.

We didn't speak one another's language, but Belyaev spoke some Spanish and German, as did I, having learned the latter when I was stationed in Germany. So we were able to converse pretty well. At dinner that night, Belyaev and I traded the wristwatches we had worn on our space flights. It was the start of a friendship that continued right up until Belyaev died unexpectedly several years later of pneumonia.

I invited the cosmonauts to join us the next morning for breakfast in my well-appointed hotel suite. (Normally when we traveled on government business we stayed in economical motels and hotels on our "per diem," but for this trip we were on orders to live it up.) Right away, continuing the brand-new tradition, Pete and Leonov traded special pens they had used on their missions. Then we sat down to talk space over eggs, bacon, and coffee. In this impromptu setting, the beginning of a working relationship between astronauts and cosmonauts began— one that is still going on to this day.

At that time the space programs of both countries were about equal. (We would begin to pass them later during the Gemini program, and by the time we got to Apollo and the lunar missions, we took the lead and never looked back.) The Soviets had much larger boosters than we did—their rockets were 40 to 50 percent more powerful than ours. As a result, they didn't have to do all the microminiaturization that we were doing— making gauges and systems more compact and multifaceted. For example, we had one gauge that we switched over and used for readings on several different systems. The Soviets, on the other hand, were using big old single-system gauges, some of them from surplus submarines.

Our new Russian friends appreciated the openness the United States practiced when it came to our space program and our willingness to admit failure or lack of total success at times. Now, one on one—or rather, two on two—the cosmonauts met our candor with some of their own.

We had heard stories about the Soviets losing men in space, and it turned out to be true. The cosmonauts told us two men died when they had a circuitry failure and lost both their cabin and spacesuit pressure on reentry. In such a loss of pressurization, death is by asphyxiation, but also the body cells erupt internally, literally tearing the body apart from the inside out. The best thing that can be said about such a fate is that it is over quickly: twenty to thirty seconds at most.

Belyaev and Leonov told us all about the space walk. Leonov's first excited words as he pushed himself through the hatch and into space: "A man is floating free in space!" The thirty-year-old lieutenant colonel in the Soviet air force, a skilled parachutist, fighter pilot, and world-class athlete, confessed to being instantly stunned with the beauty of the sight below him.

"I didn't experience fear," he said. "Only a sense of the infinite expanse and depth of the universe."

While the USSR had claimed that Leonov's twenty-four-minute space walk went without a hitch, Pete and I now heard otherwise. After he went through the airlock and stepped into space, Leonov's suit became very stiff from pressurization (the suit may have been too big for him), and he found it nearly impossible to bend his arms and legs or otherwise control his movements. In his struggles, Leonov overheated due to exertion and on his way back into the spacecraft became stuck in the airlock. He only freed himself after releasing some of the air from his suit—a dangerous and desperate move—so that he could better move his arms and legs. If he'd released too much air, he could have suffocated.

When mankind's first EVA was history, Belyaev said, his crewmate returned to the spacecraft "looking like a man who had just been reborn—a man who had just come back from another world."

It was exciting stuff to hear about, but the cosmonauts were not finished telling of their adventure. They nearly didn't make it home in one piece.

The day after the space walk when it was time for reentry, their spacecraft's automatic stabilization system failed, requiring Belyaev to take over manual control of the spacecraft—one more experience the senior cosmonaut and I shared. This took some time to work through, and as a result their retro-fire was delayed one orbit. When they finally fired their reentry sequence, Belyaev did a good job of guiding the spacecraft through the harrowing reentry, only to find that the extra orbit had thrown off their landing site by six hundred miles in snowy Siberia. (Each orbit takes the spacecraft over different terrain, which is why coming down when scheduled is crucial to a quick recovery.)

The Soviet spacecraft crashed in the thick forests of the northern Ural Mountains, wedging itself between two large fir trees. Although able to loosen the hatch, Belyaev and Leonov remained stuck inside the freezing craft all night after their heating system failed. Adding to their misery, a hungry black bear menaced them throughout the night, trying desperately to get at them.

Theirs was the first spacecraft to return with bear claw marks, Belyaev said, amid great guffaws from all four of us.

Not until a recovery crew skied into the snow-covered forest the following day were the cosmonauts released from their confines.

"Maybe water landings aren't so bad after all," I told Pete.

The cosmonauts admitted to us that persistent problems with reentry and recovery were plaguing the Soviet space program. Because they launched over land, their spacecraft had been designed for ground recoveries. Any U.S. astronaut would have preferred landing on the ground to coming down over water—especially us air force guys, but the navy guys didn't like dropping into the middle of the ocean either. Water recoveries had been more or less forced on us because there were no suitable unpopulated sites in the continental United States for land-based launches. Since we launched over water, we had to be prepared for ocean recoveries in case of an emergency soon after liftoff. The Soviet Union, with its huge amount of unpopulated land areas, had no such constraint.

The Soviets did have some problems unique to land recoveries.

Returning spacecraft from both countries swung under a main parachute and landed at approximately thirty-two feet per second. At this rate it was not possible to land on the ground without sustaining damage to the spacecraft and possible injuries to the crew. In ocean recoveries, the water at least cushioned our fall. In Mercury, we had been stuck riding the spacecraft down no matter what. In the event that a Gemini crew ever went down unexpectedly on a land surface, we would have ridden the spacecraft down to parachute altitude and ejected for a normal parachute landing, then hitchhiked to the nearest phone booth. We all carried dimes.

Since cosmonauts were forced, by design, to ride their spacecraft down to the ground, the Soviets had devised a unique way to slow its descent: once the parachute deployed, and when a one-hundred-foot wire that hung from the spacecraft touched the ground, a small rocket went off. Like a retro rocket, it fired opposite to the spacecraft's trajectory and slowed the spacecraft down—to a reasonable ten to twelve feet per second.

The shape of the Soviet spacecraft was circular, and the cosmonauts told us of problems they were having with it tumbling after reentry before the main chute was deployed. This sometimes caused the shroud of the main parachute to become entangled around the spacecraft, and they'd had a few close calls.

I asked if they used a drogue chute, and the cosmonauts said no. Drawing on a napkin, I showed them exactly how a drogue chute worked on a spacecraft and gave them tips on installation, storage, and deployment. The cosmonauts were enthusiastic.

Not long after our breakfast, a cosmonaut died when the shrouds of the main chute wrapped completely around his tumbling spacecraft, causing a terrible impact. After that, all Soviet spacecraft were equipped with drogue chutes, which looked very much like what I drew that morning on my napkin.

I met two other cosmonauts the following year when I served as the official host for their red-carpet U.S. government–paid tour of the United States. I was given twenty thousand dollars in cash, a government credit card, and had an air force turboprop Convair at my disposal. For thirty days we traveled all over the country—New York, Disneyland, the Grand Canyon, Yellowstone. One of the cosmonauts, Major General Georgiy Beregovoy, brought along his wife, Lidiya, and their seventeen-year-old son, Viktor. The other cosmonaut, Dr. Konstantin Feoktistov, who had helped design the *Soyuz* spacecraft and had made several space flights, was divorced, although that wasn't why he was a self-professed outcast in his country. It seemed he preferred Scotch to vodka.

The cosmonauts were amazed at the quality of life in America. They were very frank about their lives back home. Although they lived far better than the average Soviet citizen, they all lived in small condos on modest stipends. None had their own home, a car, or even a washing machine; they went down to the community laundry once a week. Ninety percent of their country's gross national product, they confided, went to the military. They kept telling me how lucky we had it in this country. While we didn't get into any political discussions— there was a stern-faced KGB officer along on the trip—it was clear from their comments that the cosmonauts were impressed with what they were seeing of capitalism and democracy in America.

We visited Detroit and went on a tour of General Motors. After we saw the assembly line in operation, I arranged with a GM official to let the cosmonauts drive some experimental cars. Out onto the streets of Detroit they screeched, each behind the wheel of a souped-up car, with a GM engineer in the passenger seat giving directions.

Our visitors later said that the fast rides were the highlight of the trip, proving without a doubt that cosmonauts and astronauts were much more alike than not.

From Athens, we flew to Africa.

It was the first continent I had flown over in space, and I well remembered that view from a hundred and fifty miles up— filled with rich browns and golden light.

When I stepped out of *Air Force One* in Addis Ababa, the capital of Ethiopia, my first thought was that Africa looked just as beautiful from the ground.

My very next instinct was to run like hell. A huge *lion* was eyeing me from the top of the ramp.

Then I saw that it was on a chain being held by a military guard, and a few paces away: His Majesty, the Emperor Haile Selassie, known respectfully around the world as the "King of Kings" and considered the leading ruler of Africa.

It seemed the emperor had brought along one of his prized pets, just as we might bring the family dog on a picnic.

Selassie had a big interest in space and would come to Cape Canaveral for several launches. He was a diminutive man with high cheekbones and a beard who showed the world a proud and regal bearing. Once you got to know him there was a warmth to him, and he asked unending questions about space travel and a multitude of other subjects he found fascinating. He had an enormous curiosity, and when you were engaged in conversation with him (he spoke English with a British accent) he locked on to you as if what you were saying was the most important thing he had ever heard.

His was an early voice for African independence and unity. He had been *Time*'s "Man of the Year" in 1936, the year his reign was interrupted when Ethiopia was occupied by the Italian army, forcing him into exile. The Italians surrendered to British forces six years later, and Selassie returned to lead his country. That same year he negotiated an agreement with England that confirmed Ethiopia's existence as a sovereign state, although the two countries kept close ties.

My wife, Trudy, and I and our daughters, Camala and Janita, then fourteen and thirteen, respectively, along with Pete and June Conrad and the rest of LBJ's official "astronaut party"—we traveled with diplomatic credentials and greeted heads of state with personal letters from President Johnson— were taken to the emperor's palace, a huge multilevel manor on forty well-manicured acres.

At its entrance stood two big well-dressed guards holding gold chains. At the other end of each chain was a live cheetah with a gold collar. They were gorgeous animals, and one of my daughters asked if they were tame. Selassie's son, the chief of staff of the Ethiopian Air Force, led her over to the cheetahs for her to pet them. She found them to be as tame as our German shepherd at home.

After being shown our magnificent rooms, we were given one bit of friendly advice: at night, we shouldn't try to roam the palatial grounds. The emperor had standing orders for his beloved lions, tigers, and cheetahs to be turned loose every night and allowed the run of the place. What a novel way, I thought, to make sure guests didn't stay out too late and the kids got to bed on time.

And Selassie *did* have lots of children. When it was time to meet the family, we were escorted into a room filled with thirty or forty women. These were the emperor's wives, we were told. The oldest was in her eighties and the youngest was thirteen. Selassie had sired more than a hundred offspring.

Every Wednesday morning, women from the town were invited to come to the palace and meet with the emperor. They lined up in his court, and Selassie listened with great interest, one by one, to what each had to say—complaints, suggestions, you name it. When some action was necessary, he turned to an aide and issued an order. Listening to his country's wives, mothers, and grandmothers, Selassie explained to me, was the best way for him to find out what was really happening in his country.

One day I went shopping with the family in the Addis Ababa marketplace. Ethiopia was filled with mineral deposits, and there were some beautiful rocks and gems to be found. On this excursion, we came across a booth where scantily clad women of all races were openly being sold to the highest bidder. I got the family out of there quickly.

That night at dinner, I said to Selassie, "I thought you didn't allow slavery, your highness."

"We don't," he said firmly, shaking his head.

I told him about the booth in the marketplace and he seemed shocked. The next day, the booth was gone. I did wonder why someone else hadn't mentioned it to him before.

When we took our leave several days later, Selassie invited us to return to Ethiopia. I never did, but I did see him in Florida several times when he came for some Apollo launches. Whenever I did, he greeted me like an old and dear friend.

In 1975, with opposition growing to Selassie's reign, there was a coup, leading the way to Ethiopian socialism. Haile Selassie was placed under arrest and died under questionable circumstances. He was never even given a state funeral. After his death, Ethiopia, once considered the jewel of Africa, suffered terribly, with economic collapse, regional infighting, and widespread famine.

Our next stop was Nairobi, Kenya, to visit another of Africa's powerful leaders, Jomo Kenyatta, a former author and actor in England who returned to his native Kenya after World War II and became an inspirational union organizer and political leader. In the 1950s, he had been imprisoned for seven years for his involvement in the violent Mau Mau uprising against white colonialists. He emerged from prison a popular hero and was considered the one man who could unify the warring factions of his country and provide strong international leadership for a free Kenya.

Kenya had won its independence from British rule two years before our arrival, and Jomo Kenyatta, hailed as the "Light of Kenya," had been elected its first president.

Kenyatta had made arrangements to meet our party at picturesque Kiko Rock, a hunting lodge where he and a "great white hunter" had started the Mau Mau revolution resulting in the deaths of hundreds of white settlers who had been brutal to blacks. We were to spend the day with him, then return to Nairobi.

We leased two twin Cessnas from Nairobi Airlines, meaning that Pete and I each had one to fly our families in. After all the pomp and circumstance of the royal visits, it felt good to be at the controls of a plane again. With local maps to guide the way, we soared over a great sea of green.

About the time we were wondering if we'd missed the small airstrip, a clearing appeared below. We came in low over the most amazing sight: the airfield was surrounded by waves of colorfully attired natives. We landed and found Kenyatta awaiting us.

With graying beard and hair, Mzee Jomo Kenyatta was barrel-chested and bigger than life. He held his head high and gave a strong handshake, despite his advancing years and an injured leg that necessitated his use of a cane. There was a powerful and dramatic flair to his speech and movements, as if he knew the world was watching and listening to everything he did and said.

Introductions were brief and informal.

"Tell me. If you go hunting," Kenyatta said in a deep baritone with a clipped English accent, "which would you rather carry? A gun or a camera?"

I had gone deer and bird hunting in Colorado plenty of times, but I also enjoyed taking pictures of animals in their natural habitat. I looked at Pete, but he stayed quiet. "A camera," I said.

Kenyatta eyed me as if I had given sworn testimony in court. "Right answer," he said, then grinned.

Unbeknownst to us, Kenyatta was a great fan of saving wildlife, and we had landed in the middle of a national game preserve. A presidential Land Rover was nearby, and Kenyatta, at the wheel himself, took us on a personal tour of the bush, showing us lions, giraffe, antelope, and deer in their natural settings.

Before we departed the airfield, we were asked to inspect the troops. The natives we had seen from the air—standing in row upon row with the discipline of U.S. Marines—turned out to be Masai warriors, said to be among the fiercest fighters in all of Africa. If there was one under seven feet tall, I didn't see him.

We walked the front rank with a Masai chief and an interpreter. I stopped and pivoted sharply in front of each warrior, looked them up and down, then moved on. As I stepped before one warrior, he cleared his throat and said something in his native language.

The interpreter repeated the warrior's question in clear English: "Sir, after your fuel cell problem, how did the phantom radar rendezvous turn out?"

I was dumbfounded but answered the fellow's question.

I had heard it said many times that space exploration

made the world a smaller place, but until that moment I hadn't fully appreciated just *how* small.

As we walked on, I turned to the interpreter.

"Say, can you tell me—"

"Transistor radios," the smiling officer said. "Voice of America carried your flight live in Swahili."

"OUR GERMANS ARE BETTER THAN THEIR GERMANS"

A boy of thirteen pulled a red wagon out onto a dirt lane. Six crude skyrockets were strapped to the wagon, and he lit the fuse. The small wagon shot forward and careened into the town square of Wirsitz, Germany, exploding with a loud bang. The year: 1925.

With his first "launch" behind him, Wernher von Braun had found what would become a lifelong passion. His was a rocket boy's dream come true: to grow up and build rockets with which to explore outer space.

I was close to Wernher, who was the most valuable architect of America's early space program as well as its most effective spokesman and greatest salesman during those critical years from the mid-1950s until our first lunar landing in 1969. Once asked by a newsman to define briefly what it took to conduct a space program, Wernher quipped, "Early to bed, early to rise, work like hell, and advertise!"

Wernher served as director of NASA's Marshall Space Flight Center in Huntsville, Alabama, where all the rockets for the space program were developed. I first met him in 1959, during Mercury, when I went to Huntsville to learn more about rocket boosters at the Army Redstone Arsenal, where Wernher, the lead scientific adviser, and his team of Germans had built the first U.S. ballistic missile: the army's Jupiter.

Meeting Wernher and his team gave me a healthy dose of confidence in our fledgling space program—we were lucky to have them on our side. I could see that Wernher and the other Germans took the space race very seriously, and were determined to help their new country, America, get men into orbit and beyond.

From my service in Germany in the early 1950s, I had come to know and respect the German people. I found them to be reliable, efficient, and hard workers. I sometimes wondered, in fact, how in the world we had beaten them in the war.

Wernher was brilliant, as one would expect, but I also found him to be a marvelous conversationalist, raconteur, and genial host—as well as a skilled player of the cello and piano. He was a big man who had stayed in shape; powerful shoulders, a square jaw, and a full head of dark, slicked-back hair. He spoke English with a fairly heavy accent, and I think I surprised him when I first spoke German to him (he told me I had a heavy accent). He was quick with a joke and to break into a grin, and when he did, his crystal-blue eyes sparkled with delight.

On many occasions, I sat at the bar in his home all night long talking space until the sun came shining in the windows. We shot the bull about our fantasies of space exploration—including making a trip to Mars, one of Wernher's lifelong dreams. He believed Mars offered a friendlier environment than other planets in our galaxy, and we concurred that human colonization of Mars might one day be a reality, particularly with the Earth becoming overcrowded and polluted.

Every time I left Wernher's company, I felt that I could have stayed for several more days talking about space exploration. Like me, he felt certain that the universe is teeming with life and that there are other civilized planets out there.

He regaled me with tales of his life in Germany, before and during the war. As I listened, mesmerized by his stories, I realized I was learning not only about this man's amazing life but also about the very roots of the U.S. space program—for without Wernher von Braun and his fellow German rocketmen, we would still have been tethered to Earth's gravity.

I heard how, as a youth, Wernher first became enamored with the possibilities of space exploration by reading the science fiction of Jules Verne and H. G. Wells. It was a more academic account, *By Rocket to Space,* written in 1923 by German rocket pioneer Hermann Oberth, that prompted young Wernher to study calculus and trigonometry so that he could understand the physics of rocketry.

In 1929, at age seventeen, Wernher joined the Germany Rocket Society, a group of students and young engineers and scientists who shared his vision. The society's members followed the work of an American, Rob Goddard, who shot liquid-fueled rockets high into the atmosphere. Although Goddard was generally ignored by his fellow American scientists, most of whom were shortsighted when it came to rocketry, these young Germans built on his discoveries, adding them to those of Oberth, as they designed, built, and tested increasingly sophisticated model rockets.

Wernher received his bachelor's degree from the Berlin Institute of Technology in 1932, at the age of twenty, and two years later, his doctorate at the University of Berlin. He soon joined Germany's military rocket development program. The German government had discovered a loophole in the Treaty of Versailles, which after World War I limited Germany to a small army and navy and prohibited submarines and warplanes but made no mention of rockets. Within two years, Wernher was one of the leading scientists in Germany's rocket development program, which was the only one of its kind in the world. Wernher would later admit that World War II–era rockets were expensive and largely inefficient weapons and that they were sold to the military by scientists such as himself primarily as a way to get funding for their research.

As well as being a great storyteller, Wernher had some fabulous pictures of early German rocket development stored in several albums. The scotch and bourbon flowed freely at Wernher's bar, and I listened to the tales he told. Some were inspiring and exciting. Others came from somewhere inside Wernher where he had painful memories of the war years and felt very lucky that he and his family—his wife, Maria, and three children—were able to get a new start in America. (Wernher had not been a member of the Nazi Party, and had no use for Hitler, whom he referred to as "the madman," as did many good Germans I had met.)

Through the 1930s, Wernher and his team continued to develop rockets for the German army. Under pressure from the Third Reich, his dreams of space travel were subjugated to ever-increasing demands for weapons. He was once even imprisoned for two weeks by Gestapo chief Heinrich Himmler on suspicion that the young scientist was interested in building rockets only for space flight, not for warfare. It took a direct appeal to Hitler by General Walter Dornberger, head of Germany's military rocket program, to free him.

Hitler ordered that all war-related efforts be carried out in the strictest secrecy, so the rocket-testing program was moved to an island just off the northern coast of Germany in the Baltic Sea, near the old fishing village of Peenemunde. The site was suggested by Wernher's mother, who remembered the isolated, marshy seacoast as the elder von Braun's favorite duck-hunting spot. Eventually the site would become a bustling "rocket city," housing engineers, technicians, scientists, support personnel, and manufacturing facilities.

In the beginning, Hitler had not been enthusiastic about rockets, Wernher remembered. The leader of the Third Reich thought he could win the war quickly with troops, panzer divisions, and air power without resorting to such new technology. The program languished as a backwater operation until after the Luftwaffe lost the Battle of Britain. Only then did the rocket program swing into high gear, with orders to build rockets that could fly higher than the British radar shield and bomb London.

First came the V–1, nicknamed the "Flying Bomb." Between June and September 1944, more than two thousand of the pilotless aircraft—really a rocket with stubby wings—crash-dived into London, seriously disrupting rail and transportation networks, inflicting damage on factories, and killing several thousand people.

Next came the world's first ballistic missile, the V–2, which on its maiden flight, on October 3, 1943, became the first man-made object—other than a bullet—to achieve supersonic speed. A liquid propellant missile extending some forty-seven feet in length and weighing 27,000 pounds, it reached a speed of thirty-five hundred miles per hour and an altitude of more than fifty miles. Capable of delivering a 2,200-pound warhead to a target five hundred miles away, the V–2 was first employed against targets in Europe beginning in September 1944. After a V–2 first hit London, Wernher remarked to his colleagues, "The rocket worked perfectly, except for landing on the wrong planet."

As Hitler's war machine collapsed, Berlin ordered the Peenemunde rocket team to destroy all classified documents. Wernher disobeyed, hiding them in an abandoned mine in the Harz Mountains, from which they could later be retrieved. In May 1945, with the Russian army rapidly approaching Peenemunde, the majority of the rocket team filled up gas-starved cars and trucks with methanol drained from the rockets and raced three hundred miles to Munich, where Wernher and 117 of his key team members surrendered to American forces. If they hadn't made that dash for the American lines, *all* the German rocket experts might have ended up in Russian hands—keeping the United States out of the space business for a long time. "We'd had enough of a totalitarian society," Wernher explained, "and didn't want to live in another one."

After the war, the Germans were taken to an army base at White Sands, New Mexico. There they spent their days tinkering with captured V–2 rockets and teaching rocketry to those in the army who were interested.

In 1950 von Braun's team was responsible for the first rocket launched from Cape Canaveral—a fifty-six-foot WAC Bumper, a modified V–2, which climbed ten miles above Earth. It was a modest enough beginning, with a cement slab serving as launch pad and a tarpaper shack for the control center. Wernher and his colleagues were soon moved to the Army Redstone Arsenal in Alabama, where they built this country's first ballistic missile: the Redstone. Once more the German scientists had been called upon to build a weapon of war—this time for use in Korea. Wernher was disappointed at having his hopes for space rocket research and development put on hold again. But Redstone, which was never fired in anger, would prove to be his ticket to space: it was used to plant America firmly in the space race by launching both Al Shepard and Gus Grissom on their suborbital flights.

Wernher and his army rocket team were given the job—by President Eisenhower—of helping the United States catch up after the first Russian Sputnik in October 1957. They succeeded in putting the first American space satellite, Explorer I, into orbit three months later—on January 31, 1958.

In 1960 the German scientists were transferred to the newly established NASA and received the mandate that Wernher had long wanted to hear: build a rocket capable of sending men to the Moon. His answer: the mighty Saturn rocket. I would learn that before President Kennedy announced his national goal of reaching the Moon before the end of the decade, he had asked Vice President Johnson to canvass luminaries in the scientific field, including Wernher, for their best ideas. Wernher wrote an apparently persuasive and even prophetic memo championing the cause of a manned lunar landing. Within days of reading the memo, JFK went before the nation and presented his historic goal.

Wernher and I were often joined at his home by others of the more than one hundred expatriated Germans who came to America and helped us win the space race. They were known as the "Huntsville Gang." The Soviet Union had its own ex-German scientists, and they achieved a head start because the Russians were willing to finance space research when our leaders remained uncertain of the value of space exploration. But once we assumed the lead, we never relinquished it. As we enter the new millennium, it is worth remembering that no one other than an American has set foot on the Moon or even broken away from Earth's orbit. As we always said at the time: "Our Germans are better than their Germans."

The visitors to Wernher's house included Kurt Debus, who had also been at Peenemunde and was an extremely capable rocketman. He headed up the booster launch operations for Mercury and Gemini at the Cape, and later became director of the Kennedy Space Center. Kurt was a graduate of the University of Heidelberg and a serious-looking German type, direct from central casting, complete with saber scars on his face from learning the art of sword-fighting.

Then there was Joaquin "Jack" Keutner, with whom I worked in the early days of Mercury on the Redstone rocket program. Jack had some hair-raising flying stories to tell. In an attempt to improve accuracy over the target, some V–1s were modified with a cockpit to allow for a pilot. Jack had flown several trips across the English Channel atop a V–1 strapped under a twin-engine Junkers bomber. After being dropped free, he would air-start the "Flying Bomb." When they got within range of London, he would release the bomb, then turn toward the French coast and ride the rocket home. Before landing, Jack would dump any remaining fuel and glide the V–1, modified with landing skids, to a belly landing in a field.

One time, things didn't go as planned. For this flight, they had a two-man cockpit in a V–1, and Jack was checking out a less experienced pilot. When they were dropped by the Junkers, they couldn't get an air start and had to turn back toward France. Jack released the warhead but was unable to dump fuel, so they came in heavy, loaded with combustible fuel and at a high rate of speed: in excess of 270 miles per hour. They hit the field, slid its entire length, and went into a pine forest, leaving a trail of burning debris behind them. The rocket disintegrated. Somehow Jack got out, but the other fellow didn't.

At war's end, a manned V–2 was sitting on the pad at Peenemunde, all tested out, fueled up, and ready to go. It would have been launched on a low-energy easterly orbit, Jack explained. The plan: to drop a warhead on New York City. That 1945 manned rocket flight—sixteen years before the first U.S. manned rocket flight—came within a week or so of being launched.

Wernher confided to me that the Germans were testing more than rockets at Peenemunde. "Some of the craft we were developing," he said, "were far ahead of anything the rest of the world had or knew about."

"You mean jets?" I asked, thinking of the Luftwaffe's Me262, the world's first jet fighter.

He smiled a scientist's knowing smile. "You could almost not refer to them as planes. We flew several craft that were totally different. Very advanced principles were involved."

According to Jack, who flew some of these advanced craft, they included saucer-shaped vehicles with double intakes and counter-rotating fans and disks and some with advanced propulsion systems. Jack said they had flown successfully. None of these craft surfaced after the war. Wernher and Jack were unclear about whether any of them had survived the war's last hectic days.

Early German rocket scientist Hermann Oberth also ended up in the United States. He was hired in the mid-1950s as a consultant for the U.S. Army Ballistic Missile Agency and later by NASA. Oberth, whom I met through Wernher, had a brilliant mind that accepted few limitations to man's reach. He had strong views about UFOs, having been hired by the West German government to head a commission for the study of UFOs following the war. In the commission's final report, Oberth contended that some of the unexplained objects were "propelled by distorting the gravitational field, converting gravity into usable energy." He believed there was "no doubt" that some of the unexplained objects were "interplanetary craft of some sort that do not originate in our solar system."

Another former top German scientist who agreed with Oberth was Walther Riedel, once chief designer and research director at Peenemunde. Riedel, who also came to this country to work in the U.S. space program, kept records of saucer sightings around the world. Convinced that some of the sightings had an "out-of-world basis," he offered several strong arguments to support his belief: the skin friction of the craft at speeds and altitudes observed would melt any known metals or nonmetals available; the high acceleration at which they flew and maneuvered would be intolerable for the crew; the many instances in which they had done things that only a pilot could perform but that no human pilot could stand; the fact that in most of the sightings there was no visible jet flame or trail, suggesting "no power unit we know of." From my own experiences and those I'd heard about from pilot friends, I was in complete agreement with these findings.

Wernher told of a firsthand UFO demonstration that had been given to him and other German rocket scientists and U.S.

military personnel at White Sands, New Mexico, on July 10, 1949. While tracking the test launch of a V–2 rocket at two thousand feet per second, scientists suddenly saw two small circular UFOs pacing the missile. One was seen to pass through the missile's exhaust and rejoin the other object. Then they quickly accelerated upward, leaving the missile behind. Seeing their capability for hovering around a speeding rocket and accelerating away from it with apparent ease left a vivid impression on all who witnessed the event.

I was amazed at how long Wernher and the other German scientists had been thinking about rockets and space—*decades* longer than their counterparts in this country. Without them, the United States would have been at least ten or twelve years behind the Soviets in the space race. Perhaps we would *never* have made it to the Moon.

Wernher was known to be so anxious to start exploring space that there was some concern among NASA officials about whether the brilliant German scientist might jump the gun. For the first long-range test in 1960 of the Atlas booster, which was to be fired at the Cape and land in the ocean off South America, a bunch of extra security men were assigned to the Launch Control room to make sure Wernher didn't reprogram the rocket for an orbital flight. Wernher got a big kick out of that.

Wernher was a can-do kind of guy who got things done.

I was in Huntsville one time, working on some projects in preparation for an upcoming Apollo mission. I was in charge of flight crew operations, and we needed to do some crew training for extra-vehicular activity. The best way to train a crew for venturing out into space was to put them in a tank filled with water; working underwater closely resembles zero gravity. NASA had been building a water tank for that purpose in Houston, but it had been going on for months. The latest forecast was that it was still months from completion, which would make it almost useless for the crew training we needed to do.

That evening, Wernher called me at my office. He said he'd heard about the problems we were having with the new water tank. "When do you need this tank?" he asked.

"We need it right now."

"Well, I have one you can use. It's sitting out back behind one of the buildings. Would you consider using it?"

"Is it man-rated?" I asked. We never put astronauts inside something that wasn't man-rated, meaning that someone had gone in first, checked out all systems, and found things ready to go.

"It will be man-rated tomorrow," Wernher said, hanging up.

And who man-rated it so quickly? *Dr. Wernher von Braun,* that's who. He put on a space suit and scuba gear and climbed into the water-filled tank. He called me back soon after he emerged.

"It's man-rated now," Wernher chuckled.

And that's how we finally got our tank for crew training.

In addition to finally achieving his goal of exploring outer space, Wernher became an American citizen, and a dedicated one at that. He thought the United States was the greatest place he'd ever seen. Even though in Germany he had been a top scientist with all kinds of perks and priorities, he'd been forced to live a carefully controlled life and be constantly on guard over what he said and did—things that people born free tend to take for granted every day.

I never once saw Wernher take them for granted.

As I mentioned, Wernher had a humorous side. He loved a good joke and had a quick wit. At a time when the United States was still far behind the Soviets in the space race, Wernher was asked by a newsman to defend NASA's existing program for achieving supremacy in space rather than giving in to the cries for "panic" programs. "Those type of programs will fail," Wernher said, "because they are based on the theory that with nine women pregnant, you can get a baby in one month."

In the early days, after so many rocket failures, Wernher

was asked by a reporter if it appeared that NASA needed a crash program. "What we need is less crash and more program," Wernher said.

Asked what he thought would be the first thing we found on the Moon—this after several early Soviet successes in space while the United States was still far behind—he answered, "At this rate, an empty vodka bottle."

Wernher had a plan to get an American on the Moon some three years earlier than we eventually did. Called Project Adam, his idea was to send one man to the Moon and back—a quick trip, without any Moon walks—just to get someone there safely before the end of the decade. It involved using two spacecraft— one as a fuel tanker and the other to blast off to the Moon from space. His would have been an entirely different approach from what we ended up doing: landing a two-man team on the Moon with an emphasis on conducting lunar research.

There was serious discussion inside NASA in the early 1960s about Project Adam. People in some quarters thought it was pushing the Moon landing up too soon, with increased hazards as a result. In the end it was decided to take the longer and more conservative approach. Knowing Wernher, I'm sure his plan would have worked just fine.

I would have been happy to fly his one-man mission.

On November 16, 1963, President Kennedy visited Cape Canaveral to get a firsthand look at the new launch facilities for Project Apollo and the huge Moonport under construction on adjacent Merritt Island.

Kennedy first spent time with Wernher, who took great delight in showing off his latest creation: the powerful Saturn I rocket booster, which was being readied for its first all-up test flight.

"If we make a booster that's very powerful," Wernher once said, "I will cluster four or five of them together and make it bigger." Cluster, he did—the later version, the awesome Saturn V, sported *eight engines,* which could produce 180 million horse-power. To this day, Saturn V is the most powerful rocket ever built by man.

Saturn was the rocket that would be needed to keep JFK's promise of reaching the Moon by the end of the decade, and the president knew that.

Later, Kennedy climbed into a navy helicopter with Gus Grissom and me. We gave him a first-class bird's-eye view of the new Moonport, where one day in the not-so-distant future a Saturn would sit with a manned Apollo spacecraft atop it.

Six days after his visit, Kennedy fell to an assassin's bullet in Dallas.

I spoke with him the night before he was killed. We were together at a function at Brooks Air Force Base in San Antonio, and the president came over and asked me if I could go to Dallas with him the next day. He said he could use a "space hero" with him on the trip. I couldn't make the trip because some important systems tests were scheduled at the Cape for the next day: November 22, 1963. Had I gone, I suppose I would have been riding in the presidential motorcade that day.

With Kennedy's death, a pall fell over the Cape as well as the nation. The man who had so eloquently and so brashly put us on course to the Moon didn't live to see it.

The year after the first lunar landing in 1969, NASA leadership asked Wernher to move to Washington, D.C., to head up the strategic planning effort of the agency. In that capacity, he headed the task force that proposed the space shuttle, which, three decades later, is the backbone of America's space exploration program.

In 1972 he retired from NASA and became vice president for engineering and development of Fairchild Industries in Germantown, Maryland. He was also active in establishing and promoting the National Space Institute, a nonprofit group that brought together government and industry for continued space research. At the peak of his activities, he learned he had cancer, which forced him to retire in December 1976. He died six months later in Alexandria, Virginia, at the age of sixty-five.

I am proud to have known one of the most important rocket developers and champions of space exploration of the twentieth century.

I am also most proud to have called Wernher von Braun my friend.

"THERE'S A FIRE
IN HERE!"

Gus Grissom and I had a long talk the day before he died.

The first American to go into space twice, Gus had been selected as commander of *Apollo 1*, which would have made him the first three-time space traveler.

We had our last conversation in the astronaut office at the NASA Manned Space Center in Houston, where we lived within a couple miles of each other.

Gus and I were like brothers. We bitched and moaned to each other, patted the other on the back when he needed it, worked together on souping up race cars, and generally had a ball together.

That last time I saw Gus, he wasn't his usual buoyant self. He thought his mission had a "pretty damn slim chance" of going its full fourteen days. With three weeks before launch, he was agonizing over the condition of his spacecraft. Some sixty major discrepancies had been identified, and there was no time to fix them before the "hot tests" scheduled for the next day at the Cape, which had been renamed Cape Kennedy in honor of JFK. This final test was just like launch day except that no one would flip the ignition switch on the booster. With the three astronauts onboard and suited up as if it were the real thing, the cabin would be pressurized and all systems activated to see if the spacecraft could sustain itself on its own.

The wise thing to do, Gus knew, was to put off the next day's dress rehearsal and give technicians time to fix all the problems with the spacecraft. That would probably take a couple of weeks. Another "to do" list would certainly result from the "hot tests," and additional time would have to be allocated for those fixes too, before the launch.

To cancel the "hot tests" meant that *Apollo 1* would fall behind schedule, and no one at NASA was in favor of *any kind* of delay in the schedule. Gus had continued to push for the fixes and was finally told that postponing the tests was his decision alone to make. While the commander could always scrub a mission if he felt it was unsafe, a lot of NASA higher-ups were passing the buck.

The space race with the Soviets was still very competitive, but mainly there was the deadline of landing a man on the Moon by decade's end. President Kennedy's national dream had not died with him. And at this point it was generally felt we couldn't afford a day's delay, much less two weeks. Several Apollo missions would have to be successfully flown before we could attempt a lunar landing, and it was already 1967. Pushing back the launch of *Apollo 1* would threaten the already tight schedule.

It was a terrible decision for Gus to have to make, and he was doing a lot of soul-searching. He wasn't seeking advice, and I didn't offer any. This was his call.

Gus made the call; probably the one I would have made in his position. Tragically, it was the wrong one.

The next day I was in Washington, D.C., taking part in the signing of the new Peace-in-Space Treaty, which aimed at preventing territorial or military rivalries in outer space and block-ing the orbiting of nuclear warheads, so that space would not become a future battlefield. Other astronauts were present, as well as a couple of Soviet cosmonauts—an international crowd on this historic day for space exploration.

President Lyndon Johnson described the treaty as "an inspiring moment in the history of the human race." Soviet ambassador Anatoly Dobrynin told the East Room audience: "Let us hope we shall not wait long for the solution of earthly problems."

I was wiped out, and after the ceremony I went over to the Georgetown Inn to go to bed early. I hadn't been in the room more than a few minutes when the phone rang. It was Congressman Jerry Ford, a friend and member of the House Space Committee. I had just left him on the Hill, and he knew I was heading to my room.

"Gordo, I just got word that there's been an accident at the Cape and the crew was killed."

I felt my mouth go dry.

The hot tests . . . the *sixty major discrepancies* . . .

Gus was gone, and so were Ed White and Roger Chaffee. Ed had flown in Gemini, and been the first American to walk in space. Roger never made it to space.

"Any details?" I managed to ask.

Pilots always want to know the details of an accident. It was a hard habit to break, even as I struggled with the emotions of knowing I had lost my best friend.

"There was a fire," Ford said softly.

Whatever had gone wrong, with all that oxygen onboard it was an environment that would support combustion. I had a feel-ing it must have been over pretty quickly for the guys. Maybe just enough time to say a prayer or at least start one.

The Apollo Review Board submitted its report three months later. It concluded that the *Apollo 1* atmosphere was lethal twenty-four seconds after the fire had started and that the crew had lost consciousness between fifteen and thirty seconds after the first suit failed. Suit failure took two to three minutes. "Chances of resuscitation decreased rapidly thereafter," the report stated, "and were irrevocably lost in four minutes."

Gus, Ed, and Roger had climbed into the spacecraft about 1 P.M. and were sealed inside. As they entered the spacecraft, Gus complained that it smelled funny, sort of like sour milk, although no one ever figured out where the odor came from. Pure oxygen at a pressure of 16.7 pounds per square inch was pumped into the cabin. The test went routinely during the afternoon, with the usual glitches. Once, when they were having a communication problem with the test-control sites, Gus snapped, "If I can't talk with you only five miles away, how are any of us going to talk to you from the Moon?"

At 6:31 P.M., shortly before sunset, one of the astronauts cried out over their radio, *"There's a fire in here!"* More cries and yells and curses followed.

The audio tape of the radio communications between the spacecraft and Launch Control was never released to the public. I heard it once, and it was terrible.

It took technicians five minutes to get the hatch open, and by then the fire had burned itself out. The crew was dead, not burned to death, as one might have expected, but asphyxiated by inhalation of toxic gases.

It was determined that the fire started due to a frayed wire near Gus's couch.

There was no fire extinguisher in Apollo, a tragic oversight. One top administrator at NASA was opposed to having fire extinguishers aboard our spacecraft; he thought the chances for a fire in space were minuscule and not worth the extra weight of fire extinguishers. It's easy in hindsight to say what was needed. We were all guilty of not pushing for them hard enough.

As a result of the accident, water extinguishers were placed aboard Apollo, even though we conducted tests that proved Halon extinguishers chemically rendered combustion impossible instantly. (To this day, NASA has only water, not Halon, extinguishers aboard its spacecraft—a policy that makes no sense whatsoever. I hope this misguided policy will be changed before another accident has to happen.)

There was almost a fourth person aboard *Apollo 1* for the "hot tests" that day. Flight Director Gene Kranz had considered being in the spacecraft to try to figure out a recurring problem with one of the systems. He would have been situated under one of the three crew couches in the sleeping area. That was where the smoldering fire started, and had Kranz been there, he might have detected it in time.

But Gene decided to stay in Launch Control.

I'm not sure he has ever forgiven himself for not being aboard *Apollo 1*.

Another piece of bad luck for *Apollo 1*: its hatch took fourteen different turns and other procedures to open up. In such an emergency it was practically unusable. A new hatch had already been redesigned for greater safety, and this was to have been the last time the old hatch was used. As soon as *Apollo 1* went into final "rehab" for its upcoming mission, the old hatch was scheduled to be replaced with the newer one, which could be opened within twenty seconds. The new hatch, which *Apollo 1* would have flown into space with, probably would have saved Gus, Ed, and Roger.

On a brisk wintry day, I walked beside the flag-draped coffin of my buddy, Gus, at Arlington National Cemetery in Washington. Roger was buried there later the same day; Ed at his alma mater, the U.S. Military Academy at West Point.

Gus left behind his wife, Betty, and two sons, Scott and Mark, both in their early teens. (Gus's boys grew up with his love for flying and today are commercial airline pilots.) The family stood in the front rank, Betty in black, the two boys staring at the hallowed ground before them. Standing with them were the president of the United States and other men and women high in government and the space program.

Rifle volleys split the air, and a lone bugler sounded taps.

Six of us in uniform stood rigid at attention. We had once been the proud seven, and had spent nearly a decade together facing every challenge put before us.

I looked up at four streaking air force fighters coming in low in a tight fingertip formation. Just before they reached us, the number two aircraft suddenly pulled up and away, leaving a glaring hole: the missing-man formation.

The symbolism was as old as anything I knew, yet I had never felt its meaning more than I did at that moment. I had lost my truest wingman, and that gap in the formation represented the hole I now felt inside me.

Good ole Gus was gone, laid to rest on a frosty hillside not far from the eternal flame that marked the grave site of the one man who most wanted us to make it to the Moon.

The long delay in our space program that no one wanted happened.

We faced the genuine possibility that the Apollo program might be canceled. It looked as if an earlier remark made by someone in NASA in discussing the safety of spaceflight—"We can't afford to lose astronauts. If we lose astronauts, we lose funding"—might come true. A national debate broke out over whether the country should be spending more than twenty billion dollars to send men to the Moon when there were so many other problems to address, including the war in Vietnam. The space program had always had its critics, and their ranks swelled overnight. Even some of our nation's most prominent scientists contended that less expensive and far less dangerous unmanned craft could learn just as much about the Moon as astronauts.

Other astronauts and I found ourselves being asked tough questions by reporters and politicians alike about the need for "dangerous" manned space exploration; they were getting an earful from proponents of unmanned space probes who saw this as an opportunity to increase their own funding and efforts. The point I continually made to the press, public, and politicians was that in the history of flight there had always been fatalities, and always would be. "We shouldn't shut down the program because of fatalities," I said. "We need to find out what caused it, fix it, and forge ahead." The information to be gained from manned flight, I pointed out, was more by a huge multiplicity than what could be learned from unmanned probes.

Into the growing debate entered NASA's best salesman, Wernher von Braun, who eloquently told Congress and all who would listen that landing men on the Moon would open a new frontier in space exploration. "When Charles Lindbergh made his famous first flight to Paris, I do not think that anyone believed his sole purpose was simply to get to Paris. His purpose was to demonstrate the feasibility of transoceanic air travel. He had the farsightedness to realize that the best way to demonstrate his point to the world was to select a target familiar to everyone. In the Apollo program, the Moon is our Paris."

Apollo was saved, largely due to the backing and political savvy of President Lyndon Johnson. While he had been trimming NASA's budget all along, to help pay for new domestic programs, LBJ refused to use the Apollo tragedy as an excuse to cut our funding further. Like the rest of us, President Johnson believed in our goal of landing a man on the Moon before the end of the decade. Exactly 1,341 changes would be made to the *Apollo* spacecraft before it was deemed worthy of once more attempting to fly in space, an effort that took 150,000 men and women working a year and a half.

In the aftermath of the fatal fire, NASA did a complete review of all the materials aboard the spacecraft. We never realized, for example, that aluminum would burn almost like wood in an atmosphere of pure high-pressure oxygen. All the Velcro we had inside the cabin to stick things to under zero gravity turned out to be highly flammable. We probably had enough papers in the flight plan alone to stack them twenty inches high, and of course all that paper was flammable.

As a result of the total safety study, we developed fireproof paper and Velcro, and other nonflammable materials. It's a shame that the FAA has not availed itself of that multibillion-dollar flight safety effort and required commercial airlines to use these materials that won't burn even when a blowtorch is applied to them. Today, materials and fabrics are used in our nation's airliners for seat padding, floor mats, and overhead lining that will burn and smoke easily and put out very toxic fumes within a matter of seconds.

Only after all this work was completed did Apollo finally fly.

At the Cape, a bronze plaque was dedicated: "In memory of those who made the ultimate sacrifice so others could reach for the stars. God Speed to the crew of *Apollo 1*."

I hated that Gus was gone, but what we learned from the disaster that took those three lives helped strengthen the entire Apollo program and ensure that we kept President Kennedy's promise.

I don't think Gus would have had it any other way.

REACHING THE MOON
AND LOSING MARS

In late 1968, with the U.S. and Soviet space programs back on track after the fatal accidents and lengthy investigations, the race for the Moon heated up.

The Apollo spacecraft, designed to take us to the Moon and bring us back safely—the main objective of the Apollo program—was a three-part vehicle.

The command module, where the three astronauts sat during launch and reentry, was twelve feet high and nearly thirteen feet at its base—downright roomy compared with the older Mercury and Gemini spacecraft. It weighed 12,392 pounds at launch and was constructed primarily of aluminum alloy, stainless steel, and titanium. The command module used only two thousand watts of electricity, similar to the amount used by an oven in an electric range. Its control panel display included twenty-four instruments, 566 switches, forty event indicators, and seventy-one lights.

Apollo's service module—twenty-two feet long—was like an extra room attached to the backside of the command module. It contained the electrical power system, fuel, the main propulsion engine, and other systems. The service module would accompany the command module to the Moon and back, then separate from the command module just before reentry into the Earth's atmosphere and become space junk until its orbit eventually eroded and it burned up in our atmosphere.

The third part of Apollo, the lunar module, was the first craft designed to fly in the vacuum of space. With no atmospheric resistance in space or on the Moon, there was no need for swept wings or other supersonic design characteristics meant to squeeze out every inch of performance and speed, which explains why it could have such an ungainly design, and still be effective. The lunar module was composed of two basic parts sitting atop spiderlike shock-absorbing legs. The top portion was the crew cabin; the bottom, the descent stage, which also served as launch pad when it was time for the astronauts to fire the ascent engine and leave the Moon's surface. The lunar module was twenty-three feet high and was made of aluminum alloy.

The booster for the Apollo missions was the newly built Saturn V, an extraordinary piece of hardware that would provide the propulsion for America's most ambitious space missions. Thirty-six stories tall, the power of its first stage alone was greater than that of five hundred jet fighter engines.

Stage number one lifted the 6.4 million pounds of rocket and spacecraft to an altitude of 38 miles in two and a half minutes. At that point the vehicle was moving more than 6,000 miles per hour and was fifty miles out over the Atlantic Ocean. It then shut off, and the empty first stage dropped off.

The second stage then ignited and burned for six and a half minutes, carrying the vehicle up to 108 miles and a speed of 15,500 miles per hour. When it was done, the second stage dropped off too, freeing the Saturn of more dead weight. Including spent fuel, the vehicle now weighed only 5 percent of its launch weight.

After a short burn (about two minutes) that finished insert-
ing the spacecraft into orbit, stage number three did not drop
off. Following a spin or two around the Earth, checking out all
systems, stage three was refired to propel the spacecraft toward
the Moon fast enough—about 25,000 miles per hour—to escape
the Earth's gravitational pull. "I have always dreamed of a
rocket which we could use to explore the solar system," said
Saturn's head designer, Wernher von Braun, upon its first successful test flight in 1967. "Now we have that rocket."

Following several successful unmanned Apollo launches,
the first U.S. three-man mission, *Apollo 7*, lifted off on October
11, 1968, commanded by Wally Schirra, who, instead of Gus,
became the first man to make three space flights. The mission
lasted eleven days in orbit—more man-hours than all the Soviet
space flights up to that time combined.

Meanwhile, the Russians, one month before *Apollo 7*, sent
an unmanned spacecraft around the Moon and back. And in the
following month, November, they loaded another spacecraft
with turtles, worms, and insects, launched it toward the Moon,
and brought the menagerie safely back to Earth.

Figuring that the Russians were getting close to making a
manned lunar flight, a quick adjustment was made. *Apollo 8*
was originally planned for an Earth orbit to further test the
command and service modules. Now, after the success of *Apollo
7* and with the Russians closing in on the Moon, it was decided
that *Apollo 8* should change its destination.

In December 1968, *Apollo 8* lifted off with Frank Borman,
Jim Lovell, and Bill Anders—the first humans to journey to the
Moon. As they orbited sixty miles above the lunar surface, the
images they captured on live black-and-white video were seen
by the largest audience in television history. Serving as narrator,
Lovell said he was imagining what a "lonely traveler from
another planet" would be thinking about the Earth seen for the
first time from this altitude. "I wonder whether I'd think it was

inhabited or not." Back at Houston, CapCom Mike Collins, a *Gemini 10* veteran, asked, "You don't see anybody waving, do you?" The six-day mission set the stage for a lunar landing.

In a final test of their huge new booster for a manned flight to the Moon, the Soviets suffered a setback in February 1969. During an unmanned test launch, its N–1 booster (even bigger than our Saturn V) erupted in a fiery explosion so powerful that wreckage was strewn for thirty miles.

This threw the door open for Americans to get to the Moon first. In the interest of safety, however, it was decided to conduct two more Apollo test flights.

Apollo 9 was launched on March 3, 1969, with James McDivitt, David Scott, and Russell Schweickart aboard. Five days after slipping into Earth's orbit, McDivitt and Schweickart went through the docking tunnel that linked the Apollo command module with the lunar module, sealed themselves off from Scott, and separated the two vehicles. After the first test flight of the lunar module in space—lasting six hours—McDivitt and Schweickart brought it back for a successful rendezvous and smooth docking with the command module.

Next came *Apollo 10*, on May 18, 1969, with Tom Stafford, John Young, and Gene Cernan—a real dress rehearsal for the lunar landing mission. After easing into lunar orbit, they not only conducted further testing of the lunar module but also perfected navigating around the Moon and confirmed a future landing site.

On July 3, 1969, the Soviets launched a second unmanned N–1 rocket, this time with a heavy *Soyuz* spacecraft atop it in a dress rehearsal for a manned lunar landing. Less than ten seconds after launch, a chunk of debris fell loose and broke through a liquid oxygen fuel line. The resulting explosion was equal to that of a tactical nuclear bomb. The Russians would not attempt another N–1 launch for two years.

Less than two weeks later, on July 16, 1969, *Apollo 11*, with

Neil Armstrong, Michael Collins, and Edwin "Buzz" Aldrin aboard, was launched. Four days later, Neil Armstrong stepped from the lunar module and climbed down the ladder that took him to the Moon's surface. With boots solidly planted on that surface, he spoke into his headset those immortal words sent back to Earth and heard around the world: "That's one small step for man; one giant leap for mankind."

I was in Houston, at the Manned Space Center, standing among the technicians, scientists, and other astronauts in Mission Control when the moment arrived. I had seen lots of space-related celebrations in my time, but on this occasion the utter delight and huge relief felt by all cannot be adequately expressed. It was the culmination of so much effort and work by so many, and the ultimate sacrifice by a few.

As I watched the images of man's first steps on the Moon, a big piece of me wanted to be up there at that moment. If there had been a warm booster on the pad, I would have been first in line. I wondered: *Will I ever get my chance?*

Armstrong was followed onto the Moon within minutes by crewmate Buzz Aldrin, as Michael Collins kept vigil in the command module circling in lunar orbit. Before leaving, they left a plaque implanted on the Moon's surface. It read:

HERE MEN FROM THE PLANET EARTH

FIRST SET FOOT UPON THE MOON

JULY 1969, A.D.

WE CAME IN PEACE FOR ALL MANKIND

It had been eight years since President Kennedy had promised to put a man on the Moon before the end of the decade.

We had achieved his goal with five months to spare.

• • •

As the youngest of the seven Mercury astronauts, I always felt
that I had a better chance of enjoying longevity in the space program than some of the other guys. I'd known pilots who could
keep going strong until they were over fifty, and that would give
me a good twenty years of space exploration. Anytime I was
asked, I always said I was *planning* to get to the Moon, but I
thought I'd make it to Mars.

Of all the planets in the solar system, Mars, the fourth
planet from the Sun, was thought to be the only place other
than Earth considered suitable for human habitation. The other
planets were either too gaseous, with little solid surface, too
windy, or too hot. Like Earth, Mars has two poles and gravity—
about 38 percent of Earth's gravity. Scientists consider Mars the
closest planet to us in terms of what we might have been once,
and agree that it's the first significant step to exploring the solar
system as well as finding out how Earth was formed. Reaching
the Moon was fine; the more difficult but more promising goal
of a manned landing on Mars had always been the dream of
Wernher von Braun, and many others of us.

In 1964 NASA had succeeded in making the world's first-
ever Mars flyby with Mariner 4. We all hoped it was a prelude to
bigger things to come.

But Mars was lost to our generation when the manned
mission, originally planned back in the mid-1960s for a 1981
launch—which, as the youngest Mercury astronaut, I believed I
was in a good position to command—was canceled. It was
killed largely at the behest of U.S. Senator William Proxmire of
Wisconsin, an influential member of the Budget and Space
Committees known for his "Golden Fleece Awards," which he
bestowed with great fanfare upon programs and projects he
thought represented "wasteful and ridiculous" use of taxpayers'
money.

At the time the Mars program was killed, the spacecraft for
the mission—about the size of the Apollo command module

with the service module attached—was 90 percent designed. We had also done a good percentage of the engineering for the flight, although it hadn't been decided whether three or four astronauts would make the trip. Importantly, we had the booster to take us there: the Nuclear Nerva. It had been built and ground-tested, but never flown in space. For the Mars mission, the Saturn would have been used to get us into Earth orbit, where we would rendezvous with the Nuclear Nerva engines and its tanks of fuel, which would have been launched separately. The technology was in place to do it. The round-trip to Mars would have taken a little more than a year. The plan was to spend a few weeks exploring the surface and conducting experiments. Rocks were to be gathered, loaded onboard, and melted down to use as solid propellant for the return home.

Unfortunately, canceling the Mars mission wasn't enough for Proxmire: he successfully pushed for an end to the Saturn rocket program. He believed this country had no business exploring outer space, and given his important position on key committees he was a tough obstacle for supporters of the space program to overcome. It was an example of what can happen when one politician gets too much power, even in a democracy. To ensure that neither NASA nor any future administration had the option to reschedule a Mars mission or even a return trip to the Moon, Proxmire saw to it that the entire Saturn V production and assembly line was shut down in the early 1970s, requiring even the destruction of the machinery and tooling necessary to build the rocket. Tragically, the technology was demolished instead of trying to work within existing budget constraints and giving consideration to possible cost-cutting measures, such as limiting NASA to building one Saturn V a year for an annual lunar mission.

In his grief over the destruction of his biggest and best rocket, Wernher von Braun, who lobbied Congress hard for a reprieve, told me in one of our last conversations that he considered it among the stupidest things this country—which he dearly loved and I'd never before heard him criticize—had ever done. I agreed. Proxmire's actions were unbelievably short-sighted, and un-American to boot.

Top NASA administrators of the time must shoulder some of the blame, for not fighting harder or more effectively against the political tide. Why would any forward-thinking nation actually *destroy* its own leading-edge technology? Killing the Saturn program was not only a waste of billions of dollars—qualifying Proxmire for a *lifetime* Golden Fleece Award—but a crime against space exploration and the advancement of technology in general. I'm *still* angry about it and will be until my dying day.

The destruction of our country's most powerful rocket hurt our national interests and continues to handicap us today. We still don't have a booster anywhere near as large as Saturn V. If we were still producing Saturn V rockets, the United States could be profiting from launching big, expensive commercial satellites into orbit, rather than being severely restricted by the payload limits of today's less powerful rockets. And of course, we would have had the boosters to do something we should have done long ago: return to the Moon and establish a permanent lunar base from which to launch future missions into deep space.

By the time of the lunar landings, Al Shepard and Deke Slayton were running the show as far as flight crew assignments were concerned. While Al and Deke were not yet on NASA flight status, they were trying hard to get into space: Al had undergone an experimental ear operation in an attempt to cure his inner-ear problem, and Deke was taking prescription drug therapy for his heart murmur. Giving the two grounded astronauts these added responsibilities had seemed like a good idea to some administrators. The space program was getting larger—with more than thirty astronauts, many of them now civilians and some with only limited flight experience. (Some newly selected astronauts had actually been sent to flight school to learn how to fly.) The handwriting was on the wall: the space program was evolving from a flight program into a scientific research program; military test pilots were being phased out in favor of civilians and scientists with doctorates from Yale and MIT.

One afternoon at the Manned Space Center, Al and Deke broke the news to me that they intended to name me command pilot of the *Apollo 13* backup crew.

I was coming off two earlier backup assignments—most recently, *Apollo 10,* which passed over the Moon at fifty thousand feet, serving as the dress rehearsal for the lunar landing, and before that *Gemini 12,* the last two-man (Jim Lovell and Buzz Aldrin) mission, which had a challenging schedule of several rendezvous and dockings, capped off by the longest space walk in history. I didn't actually fly either mission, because the regular crews remained intact and were well prepared for the flights when launch time arrived. The backup crew shadowed the prime crew, doing all the same work and readying for the mission, but in the end inevitably watched the action from the ground, while serving as Capsule Communicators, since we were the most up to speed on the procedures and knowledge of the crew.

In only one instance in America's manned space flight program did a backup crew fly the actual mission, and that was in June 1966. The prime crew, Elliot See and Charles Bassett II, were flying a two-seater T–38 from Houston to visit McDonnel Douglas in St. Louis in February 1966. Upon arrival, they lost sight of the runway in heavy snow, and See elected to circle to the left to make another pass while trying to keep the field in view. Without warning, the T–38 hit a radio antenna and plunged into a large hangar—the very building in which their Gemini spacecraft was being readied. Both See and Bassett were killed instantly. Amazingly, no one else was hurt and their spacecraft was undamaged. (See and Bassett were not the first astronauts to die. Theodore Freeman, an air force captain who never had the opportunity to fly in space, had died in October 1964 at Ellington Air Force Base outside Houston in the crash of a T–38 jet.) The mission See and Bassett were training for was flown four months later by their backup crew, Thomas Stafford and Eugene Cernan, which is why the flight carried the suffix *A*: *Gemini 9-A.*

I had completed several other ground-related assignments, including heading up flight crew operations for Skylab, America's first orbiting space station, scheduled for its maiden mission in 1973, and Apollo, for which anything involving the crew or flight operations landed on my desk first. I had also directed the Shuttle Operations design input section—presenting any changes the crews wanted in the mission or its hardware. I had so missed flying that I was getting increasingly more kicks out of boat and auto racing. In 1968 my partner, Charles Buckley, the chief of security for NASA, and I had qualified for the twenty-four-hour endurance race at Daytona. The night before the race, NASA forbade me from driving because of the dangers involved. I had no choice but to go along with them if I wanted to retain my flight status. Later NASA would amend the rules to prohibit racing only by astronauts with *upcoming flights,* but it was too late for me. At the time, asked by a reporter why we pulled out of the race at the last minute, I said with some bitterness, "I guess NASA wants astronauts to be tiddlywinks players."

After two backup assignments in a row, I could normally have expected to receive a prime crew assignment next. I had been greatly counting on getting *Apollo 13,* scheduled for spring 1970 and planned as the third lunar-landing mission.

Al and Deke were giving me *Apollo 13* all right—but as *backup.* I was furious, and told them hell would freeze over before I took another backup assignment. They just shrugged.

I knew that politics was now playing a big role in crew

assignments. In fact, to some degree it probably always had. The selection of Al, Gus, and John to be America's first three men in space probably had as much to do with giving representation to three branches of the service—navy, air force, and marines, respectively—as it did with their piloting skills.

I also feel strongly that the choice of Neil Armstrong to be the first American to set foot on the Moon was greatly influenced by the fact that he was a civilian pilot for NASA. I don't say this to denigrate Armstrong's flying skills—he was a sharp pilot and had flown the X–15. But I believe that at some level it was decided that a civilian should be the first man on the Moon. NASA later paid dearly for its selection. After his walk on the Moon—recently picked by prominent journalists and scholars as the second biggest story of the twentieth century, outranked only by the dropping of atomic bombs on Japan to end World War II—Neil came home, sat for a news conference or two, then quit NASA and became a recluse rather than take part in NASA's grand plan to milk the event for all the public goodwill possible. I think the next time Neil took a question at a press conference about his historic mission was at the thirtieth reunion of the flight, in July 1999. In this regard, Armstrong was the opposite of John Glenn, who, come to think of it, would have made a *great* first man on the Moon.

At any rate, my chances of flying with my backup crew had been sinking since Al and Deke had assigned Donn Eisele to my *Apollo 10* backup crew. Eisele was an air force pilot and veteran of *Apollo 7,* the first U.S. three-man mission. He was a qualified pilot, but he was going through a divorce, and it was well known, though never spoken of publicly, that no astronaut involved in a divorce would get a space flight. Not only was it considered bad PR but there was a concern that marital and/or psychological stress might lead to pilot error. (As it turned out, Eisele didn't return to space. After retiring from NASA, he died of a heart attack at the age of fifty-seven.)

A basic ground rule that had always been followed was to keep crews intact. Once you learned how to work well together, you knew what everyone was going to do without even thinking about it. The likelihood that Eisele would never fly again had an impact on the other members of his crew.

Clearly, Al and Deke wanted Apollo missions for themselves, and it served their own interests if crews broke up before being assigned to a prime mission because it would open up room for them once they had medical clearance—even if it openly violated the ground rules. I could see that all bets were off; things were changing and not for the better. Were they icing me by assigning me another backup role, hoping to provide more prime-crew openings for themselves in future missions?

When I confronted them with my suspicions, they didn't deny it. "Deke and I are making crew assignments now," Al said bluntly.

Deke nodded.

They were clearly offering me a take-it-or-leave-it proposition.

I had always been willing to do whatever was needed for the good of the team. My priority was always the space program itself. Delighted to be an astronaut, I had been completely trusting, believing that any flight assignment would be done with a lot of thought as to the strength and value of each individual selected. It was now a terrible blow to discover how selection was really being done these days—that personal ambition and greed were the principal driving forces. What a tragedy to happen to a program that was so important to the country and the world.

I went to see Dr. Bob Gilruth, who during Mercury had been "king of the hill" and closely involved in all the crew assignments. He said he was sorry but there was nothing he could do. I realized with a sinking heart that Gilruth, a grand old man whom I had always greatly admired and who didn't have a ruthless bone in his body, had stepped aside on his way to retirement and was now allowing Al and Deke to make the assignments on their own. I could see that I had little recourse.

Putting a couple of frustrated astronauts—with a total flight time of fifteen minutes in space between them—in charge of crew assignments had been like placing a couple of hungry tomcats in charge of the aviary.

Many years later, Al would confirm the situation while addressing a group of astronauts and friends at a dinner where he was being roasted. "At that point," he said fondly of circa 1969–70, "Deke and I had total uncontested control of crew assignments."

As I suspected, Al soon found a spot for himself in the Apollo flight schedule. He flew as command pilot of *Apollo 14*, which, after the near-disastrous *Apollo 13* flight, turned out to be the third lunar landing mission, and he got to walk on the surface of the Moon. Typically, Al had worked the system in his favor, managing to get reinstated to flight status by NASA even though at the time of his mission to the Moon the navy had still not cleared him to fly jets.

Deke made it into space in 1975 aboard *Apollo-Soyuz*— with astronauts Vance Brand and Thomas Stafford—the first international space rendezvous between the United States and the Soviet Union. (Along on that mission, as the commander of *Soyuz*, was cosmonaut Aleksey Leonov, the first man to walk in space, whom Pete Conrad and I had met in Greece.)

I wasn't the only one who considered Al a cutthroat when it came to getting his way. When he was backup on my Mercury flight, Wally Schirra—like Al, a navy pilot—volunteered to shadow Al to make sure I got a fair shake during all our pre-flight work and see that Al didn't try to sabotage my chances for taking the flight. Not that we thought he would have done any-thing to endanger another astronaut—it was just that we knew Al would do whatever he could to take someone's spot on the flight schedule. For that reason, everyone endeavored to keep one eye open in the back of his head whenever Al was around. I knew how devastated Al had been when the Mercury program ended with my flight. Had there been another mission, as originally planned, *Mercury 10* would have been his. (This predated his inner ear problem.) I think he'd felt genuinely cheated ever since.

So I wasn't much surprised by Al's shenanigans. What saddened me was that Deke, a fellow air force pilot, went along with screwing one of his own group. The pull for a top pilot to get into space was that powerful. It took me years to forgive Al and Deke for what I saw as their unfairness, but after all the years of trials and tribulations we'd been through, I still loved them like brothers and finally came to accept what they had done—although it still rankled.

At the time of our confrontation over my latest flight assignment, the three of us were the only Mercury astronauts still with NASA. Wally Schirra, the only astronaut to fly in all three of America's space programs—Mercury, Gemini, and Apollo—had left not long before, after a similar dispute with Al and Deke over the new politics of crew assignments. Believing that I would probably never get another prime crew assignment, it seemed time for me to leave too.

I had already lost Mars, and now I lost the Moon.

Shortly after my realization that I would be leaving NASA, I found myself with a decision to make: whether to retire from the air force at the same time or stay on active duty.

I went to see Air Force chief of staff General Curtis LeMay, who promised me advancement and a choice command if I stayed. "You'll get your first star right away," LeMay vowed.

General Cooper had a nice ring to it, but I had a serious concern. "Sir, I understand there's a regulation that general officers can't fly single-seat fighters."

"That's correct," LeMay said.

The reasoning was that generals didn't have time to stay proficient in the air, and the air force didn't want them going out and crashing. But I had no interest in being relegated to flying only two-seat fighters with another pilot along.

"Will you grant me an exception, sir?" I asked.

"No, Colonel. I can't grant any exceptions," LeMay answered.

With that, my decision wasn't a close call. Since I qualified for retirement with twenty-three years of service, I went out a full-bird colonel at the same time I left NASA in early 1970.

In the eleven years I'd been with NASA, I had traveled in space twice and had a lot of other adventures along the way. I wouldn't trade any of them. Becoming an astronaut had opened up a new world to the scrawny kid from Shawnee, Oklahoma— one that I was prepared for as a pilot but for which none of us were fully ready.

It had never been done before. We seven were the first, and no one would be the first to sit atop rockets and fly into space again. Our feats sowed seeds in imaginations everywhere, ours included. As former Mercury flight director Gene Kranz once reminisced: "It's almost a miracle we even got off the ground."

But miracles happened.

It was now time to see what I was going to do with the rest of my life.

11
STUMBLING
ACROSS HISTORY

During my final year with NASA, I became involved in a different kind of adventure: undersea treasure hunting in Mexico. I was a partner in the privately financed venture and helped raise money for the expedition. It was well organized—until things started to go wrong. With that, I had lots of experience.

Archives in Spain had been researched and descriptions found of three midseventeeth-century treasure-laden Spanish galleons that had been tracking northward in the Gulf of Mexico, heading up the eastern coast of Mexico, when several British men-of-war descended upon them. Desperate to avoid attack, the galleons headed inland and eventually steered up a water passageway that their charts showed as an entrance to a large inland body of water.

The passageway narrowed abruptly, but the galleons continued on, unable or unwilling to turn around. In the nick of time they came across an inlet. All three galleons turned into it, but the water was too shallow for their nineteen-foot draft, and they ripped out their wooden bottoms. As the galleons went down, the horrified crews found themselves thrashing about in the water with hungry sharks. It became a shark feed. Only a couple dozen crewmen survived, and of those, only a handful made it back to Spain to recount their nightmare.

The crown dispatched a search team to the location in an effort to recover the loot that had gone down in the shallows, but every time divers went into the water they were attacked by the sharks. The team gave up and returned empty-handed.

Old charts showed where the galleons had gone down, but three hundred years can change things a lot. What had once been shallows could now be thirty feet of water, or even high and dry. There was only one way to find out, and that was to mount an expedition.

That's how I found myself in a rented Cessna with a hired pilot at the controls, zooming low over an island set in the middle of a large bay. Our team—a dozen certified divers with scuba gear—had gotten some good magnetometer readings around a nearby peninsula, meaning that there was metal of some kind scattered throughout the area. The charts showed that the bay was in the vicinity of where the shallow inlet had once been. The plan was to hire the plane, land a couple of guys on the island with an inflatable raft, and nose around. I volunteered for the flight, along with a *National Geographic* photographer named Otis Imboden.

We pointed out where we wanted to land, but when we set down, the plane started sinking in soft sand. The Mexican pilot started to panic. I tried to calm him down, explaining that we'd unload our equipment and make a firm footprint for the plane. When we had removed most of our stuff, he revved up the engine and took off. We had pointed out firmer ground where he could land, but instead he hightailed it home.

Among the gear we'd left in the plane were shirts, canteens, wallets, hats, and sunglasses. It was bloody hot, and there wasn't a stick of shade.

We had come there to find millions of dollars worth of gold, but within no time there was something we wanted even more: *a cold drink*. We started walking, and we walked and walked—probably for a couple of hours, but it felt like half the day.

The first person we came across was an old man on a little burro. I told him we were very thirsty and asked where we could get a drink. He said there was a cantina straight down the trail. We continued on until we came to it at the water's edge. It was about the size of a small room, but it had seats in the shade and cold drinks.

The photographer and I pooled our resources: we had thirty-five cents between us. It was enough to buy two big orange soda pops. We sat there drinking our sodas, and boy, they were the best thing either of us had ever tasted.

I asked the lady running the cantina if she knew about any sunken ships and recovered treasure in the neighborhood.

"Oh, much treasure, señor," she said.

She claimed that there used to be beautiful gold statues and all kinds of other valuables right on the beach.

When we drained our sodas, we went for a walk on the beach.

As we strolled along, we found arrowheads and blades made out of a flintlike black stone. We began to fill our pockets. As we were doing so, I looked up and saw this guy's head sticking up over the top of the jungle, watching us from some vantage point.

I waved to him, and he waved back. Pretty soon we were yelling friendly greetings back and forth in Spanish. He beckoned us into the jungle, and we went inland. He turned out to be a middle-aged Indian farmer with rudimentary tools at his side. We found him standing on one of two big dirt mounds, hoeing the ground and planting corn.

The mounds were pyramid-shaped and huge: fifty to sixty feet high and forty yards wide. They had once been rock formations that had long ago decayed. Looking closer, we realized that the mounds represented some kind of old ruins.

And that's how we made our big discovery—completely by accident.

Other members of our team had become concerned about us and sent a boat out looking for us. They finally found us just before nightfall. We brought our colleagues back to the ruins the next day and dug around the base of the mounds. We found more artifacts and also some bone fragments, all of which we bagged. We took them back to Texas with us and had them carbon-tested at a private lab, along with the arrowheads and blades from the beach. They turned out to be five thousand years old!

We had planned to return to the site with some heavy hydraulic equipment and take those mounds down to nothing. Who knew what kinds of valuables there were to uncover? We all saw huge dollar signs dancing before our eyes. We would just load an airplane or two and depart with our rich bounty—as in finders keepers. But when we learned the age of the artifacts, we realized that what we'd found had nothing to do with seventeenth-century Spain. And we knew we couldn't destroy ancient history.

I contacted the Mexican government and was put in touch with the head of the national archeology department, Pablo Bush Romero. I told him I'd meet him in the coastal town of Tampico in two weeks, that I had something important to show him.

Together we went back to the ruins, which the government knew nothing about. The Mexican government ended up putting some funding into the archeological dig. The age of the ruins was confirmed: 3000 B.C. Compared with other advanced civilizations, relatively little was known about this one—called the Olmec. They lived in a large area bounded on the east by the Tuxtla Mountains and by the Sierra Madre Mountains on the south, stretching to the Gulf of Mexico. It was a humid terrain with an abundance of water: lakes, rivers, and marshes. Olmec artifacts had been discovered only within the past century, the majority as recently as the midtwentieth century. Even so, the Olmecs are considered the mother culture of civilization in Mesoamerica.

Among other things, the Olmecs have been credited with developing writing in Mexico. They also developed the concept of zero and positional numbers three thousand years before Europe did. They greatly advanced agricultural practices, which allowed them to produce high yields from small areas and feed their urban centers. They were known for building great public works.

The Olmecs were the first of four great civilizations arising in the Americas that rivaled Greece and Rome. The Olmec civilization lasted about five hundred years longer than America has been around. Olmec political, social, religious, and economic characteristics laid the groundwork for the three pre-Columbian civilizations that followed: Maya, Toltec, and Aztec.

A lot of hieroglyphics were found at our site. They closely resembled Egyptian hieroglyphics, but they were Olmec syllabic signs used to make pictures. The Olmec sculpture found at the site was divided between representations of supernatural beings and of humanoids. Skilled engineers, the Olmec had managed to transport huge blocks of basalt and other stone from quarries more than fifty miles away for their sculpted monuments.

Engineers, farmers, artisans, and traders, the Olmec had a remarkable civilization. But it is still not known where they originated, or the identity of their predecessors.

I was given a few little artifacts to keep, although most of what was unearthed rightly ended up on display at universities and museums in Mexico.

Among the findings that intrigued me the most: celestial navigation symbols and formulas that, when translated, turned

out to be mathematical formulas used to this day for navigation; and accurate drawings of constellations, some of which
wouldn't be officially "discovered" until the age of modern tele-
scopes.

I was aware that archeologists the world over had discovered places, artifacts, and written records that defied rational
explanation. This naturally led to speculation that certain
ancient mysteries might be attributed to ancient astronauts
from another world. Some of the questions were as old as,
"Who *really* built Stonehenge?"

The Olmec had used the same means of celestial navigation as the Egyptians and the Minoan civilization on Crete, and
at the same time. The navigation stars used by those civilizations are still in existence. In fact, the same stars were used by
Apollo to navigate to the Moon and back.

This left me wondering: *Why have celestial navigation signs
if they weren't navigating celestially?* Did this advanced navigational knowledge develop independently in three different
places in the ancient world at the very same time? If not, then
how did it get from Egypt to Crete to Mexico? And if so, reason
dictates they must have had help.

If they did have help, from whom?

UFOS AT THE
UNITED NATIONS

Throughout the 1970s, UFO sightings increased annually.

Countless visual sightings and radar trackings of UFOs in the vicinity of U.S. military installations were reported. In the fall of 1975, high-security bases along the U.S.-Canadian border were the scene of numerous intrusions, resulting in a report issued by the commander-in-chief of the North American Aerospace Defense Command (NORAD), which omitted one important detail: many of the UFOs were sighted over and near areas used for the storage and delivery of nuclear weapons.

From a declassified Strategic Air Command message about the UFOs: "Several recent sightings of unidentified objects over Priority A restricted areas during the hours of darkness have prompted the implementation of security Option 3 at our northern tier bases. Sightings have occurred at Loring AFB, Wurtsmith AFB, and most recently, at Malmstrom AFB. All attempts to identify these aircraft have met with negative results."

The rash of 1975 sightings included:

- October 28, Loring Air Force Base, Maine: Unknown craft with a white flashing light and an amber and orange light. Red and orange object, about four car-lengths long. Moving in jerky motions, stopped and hovered. The object looked as if all the colors were blended together. The object was solid.

- October 30, Wurtsmith Air Force Base, Michigan. One light pointing downward and two red lights near the rear. Hovered and moved up and down in an erratic manner. A KC–135 aerial tanker crew established visual and radar contact with UFO. "Each time we attempted to close on the object, it would speed away from us. Finally, we turned back in the direction of the UFO and it really took off . . . speeding away from us doing approximately twelve hundred miles per hour."

- November 7, Malstrom Air Force Base, Montana. A Sabotage Alert Team described seeing a brightly glowing football-field-sized orange disk that illuminated the Minuteman ICBM missile site. As F–106 jet interceptors approached, the UFO took off straight up, NORAD radar tracking it to an altitude of two hundred thousand feet.

- November 8, Malstrom Air Force Base, Montana. Radar showed up to seven objects at 9,500 to 15,000 feet. Ground witnesses reported lights and the sound of jet engines, but radar showed objects flying at only ten miles per hour.

- November 10, Minot Air Force Base, North Dakota. A bright, noiseless object about the size of a car buzzed the base at two thousand feet.

Even U.S. presidents were seeing UFOs.

Jimmy Carter filed a formal report while he was governor of Georgia describing his sighting of a UFO. He was leaving a Lions Club in Leary, Georgia, when he noticed an object in the sky. A red-and-green "glowing orb" hurtled across the southwestern skies that evening in January 1969, vanishing a few minutes later. Carter told his story repeatedly in the years following, becoming the first major politician to risk being labeled a "UFO crackpot." Carter told the Southern Governors Conference a few years later: "I don't laugh at people anymore when they say they've seen UFOs. I've seen one myself."

Ronald Reagan also had a UFO encounter, which I heard about from a good friend, Bill Paynter, a former air force pilot with about 45,000 hours of flying time in everything from fighters to bombers. Paynter provided the airplane that the state leased for Reagan when he was governor, and came along with it as chief pilot.

One evening en route to Los Angeles, a UFO pulled up alongside the plane and sat off their wing. Governor Reagan and the rest of his party looked out their windows at the object while Paynter did a number of cautious maneuvers to try to lose the saucer-shaped craft, which was uncomfortably close to the chief executive's plane. The UFO maneuvered with them for several minutes before darting out of sight.

Toward the end of his second presidential term, Reagan publicly speculated about what would happen if Earth were invaded by "a power from outer space, from another planet," suggesting that such an alien threat would "unite the nations of the world in a common defense." Was Reagan remembering that UFO off his wingtip?

U.S. Senator Barry Goldwater, a longtime pilot and air force reserve general, had his own encounter with a UFO one day over the Arizona desert. When he was governor, he used to fly an F–86, a single-seat fighter, on government and personal business—as governor he was answerable only to himself for the use of National Guard aircraft. During the space program, I had several informal meetings with Goldwater. At one of them, he told me about his UFO experience.

"I chased it all over the desert and couldn't get near it," he said. "Damnedest thing I ever saw. Made me a believer."

Goldwater described the UFO he played tag with that day as a "shiny saucer." It was capable of much greater speeds than his F–86, he said, and was able to stop on a dime and make turns that seemed to defy gravity.

I told the senator about my experiences with UFOs.

We agreed it was hard to believe that Earth was the only planet in the universe with intelligent life on it.

With the increase in UFO sightings, I began to wonder which influential world organization might best conduct a scientific investigation, since our own government wouldn't undertake a serious study of UFOs and go public with what it learned.

The answer, some of us hoped: the United Nations. It seemed like a good choice to serve as a central location—neutral ground, if you will—for the collection and correlation of incident reports and eyewitness accounts relating to UFO sightings around the world.

The UN agreed in 1978 to hold a hearing on the subject, and I was invited to testify about my own UFO experiences in Europe, which I had told numerous people about through the years. Other witnesses included some of the most respected UFO researchers in the world, among them Dr. J. Allen Hynek, a well-known astronomer who had served as the scientific consultant for the air force's Project Blue Book.

I would be going public, *very* public, about my belief in the existence of UFOs, and I was ready to do so. I knew that the government was keeping a lot of secrets when it came to UFOs. I felt it was high time for the veil to be lifted.

From my association with aviation and space, I had a pretty good idea of what kinds of craft existed on this planet and their performance capabilities. I believed that at least some of the UFOs—the truly unexplained ones—could be from another technologically advanced civilization, and I wasn't afraid to say so.

Before testifying, I met Dr. Hynek, the chairman of the Northwestern University Astronomy Department. It was clear how deep his commitment ran to seeking a high-caliber scientific inquiry into UFOs. The former air force consultant was now an open critic of Project Blue Book, calling it a "nonstudy" and the air force statistics on UFOs a "travesty."

According to Hynek, the air force never devoted enough money, manpower, or attention to the problem of UFOs to get to the bottom of the most puzzling cases. The military head of Blue Book was never higher than a captain, Hynek stated, and there weren't more than a handful of staff personnel to handle thousands of reports annually.

"I wasn't allowed to go through the files and look at the reports myself," Hynek explained. "They gave me the reports they wanted me to review. And of course they were under no obligation to agree with my findings."

I concurred with Hynek's assessment of Blue Book and agreed that the air force had used it more to placate the public than for anything else. I would go even further and label it a cover-up and a whitewash. Call Blue Book what you will, but while continuing to deny the existence of UFOs and eventually terminating its own investigation, the air force never attempted to offer any credible explanation for sightings of objects flying around our airspace whose flight characteristics seemed to preclude them from belonging to our military or any military on the face of the Earth.

I wasn't of the school that felt we should make a huge effort to go back historically and correlate all the old sightings. I thought we'd find out more by looking more carefully at *new* UFO sightings and carrying a coordinated and unbiased analysis into the future. The best approach, I thought, might prove to be some type of rapid response team to go into the area of a verified and credible sighting and gather information and evidence while it was fresh.

Hynek's job for the air force had been to study UFO reports and decide if their descriptions suggested known astronomical objects. Had someone reported the planet Venus or a meteor? He began to notice that many of the reports weren't made by "kooks and crackpots," as he had admittedly suspected when he first went to work for Project Blue Book, but came from credible military and civilian witnesses.

Hynek had been trying to get the United Nations to take on the UFO phenomenon for years, in the 1960s even appealing to then Secretary-General U Thant for a public hearing. But UN protocol demanded that a member nation first bring up a subject in the General Assembly before any action on that subject would be initiated.

It had taken a decade, and hundreds more UFO sightings, for that to happen.

"I am delighted to have been invited by the Grenada Mission to speak on behalf of many of my scientific colleagues about the subject of Unidentified Flying Objects," Hynek told the special panel chaired by UN Secretary-General Kurt Waldheim. "One of the smallest nations on Earth, Grenada has courageously introduced the perplexing subject of UFOs to the General Assembly . . . and trod where mightier nations have feared to tread.

"There exists today a world-wide phenomenon," Hynek went on, "the scope and extent of which is not generally recognized. It is a phenomenon so strange and foreign to our daily terrestrial mode of thought that it is frequently met by ridicule and derision by persons and organizations unacquainted with the facts. Yet, the phenomenon persists; it has not faded away as many of us expected it would when, years ago, we regarded it as a passing fad or whimsy. Instead, it has touched on the lives of an increasing number of people around the world.

"I refer to the phenomenon of UFOs, which I should like to define here simply as *any aerial or surface sighting, or instrument recording such as radar or photography, which remains unexplained by conventional methods even after competent examination by qualified persons.*"

Hynek emphasized that UFO reports had been made in "significant numbers by highly responsible persons . . . astronauts, radar experts, military and commercial pilots, many of these officials of governments, and scientists, including astronomers."

He explained that over 1,300 "physical trace" cases were then on record—"perhaps the hardest data we possess so far, the so-called Close Encounters of the Second Kind, where there appears physical evidence of the immediate presence of a UFO." He explained that this took the form of immediate physical effects on animate or inanimate matter: "Physiological effects on humans, animals, and plants have been very reliably reported, as have the interference with electrical systems in the immediate vicinity and the appearance of disturbed regions on the ground also in the immediate vicinity of the reported UFO event."

Hynek said he was not alone in the scientific world in his belief that the UFO phenomenon, "whatever its origin may turn out to be, is eminently worthy of study." He referred to a growing community of scientists from many countries who had declared an interest, either privately or openly, in pursuing the subject.

Hynek noted the need for something resembling the World Health Organization or the World Meteorological Organization through which UFO researchers could pool and share the results of their efforts. He implored the United Nations to establish a means for scientists and other specialists to work together and exchange ideas and investigative work with colleagues around the world.

"It is my considered opinion, as a scientist who has devoted many years to its study, that the UFO phenomenon is real and not the creation of disturbed minds, and that it has both grave and important implications for science and for the political and social well-being of the peoples of this Earth," said Hynek.

He looked up from his notes and admitted that he had not always held the opinion that UFOs were worthy of serious scientific study. He explained that he began his work as scientific consultant to the U.S. Air Force as an "open skeptic," in the firm belief that "we were dealing with a mental aberration and a public nuisance. Only in the face of stubborn facts and data was I forced to change my opinion."

With passion rising in his voice, he went on: "The history of science abounds with unlooked for benefits resulting from the investigation of the unknown. Who can tell what benefits might accrue from the study of UFOs? It might well lead to the solution of many pressing problems facing mankind today."

Hynek's testimony impressed me, coming as it did from such an eminent voice in the field of science.

Then it was my turn. Seated at a long table with a microphone, I began by stating the essence of my beliefs about UFOs: "I believe that extraterrestrial vehicles and their crews are visiting this planet from other planets, which obviously are a little more technically advanced than we are on Earth."

I stressed the need for a high-priority, coordinated program to scientifically collect and analyze data from all over the Earth concerning any type of encounter and to determine how best to interface with these visitors in a friendly fashion.

"We may first have to show that we have learned to resolve our problems by peaceful means, rather than warfare, before we are accepted as fully qualified universal team members. This acceptance would have tremendous possibilities of advancing our world in all areas. Certainly then it would seem that the UN has a vested interest in handling this subject properly and expeditiously."

I pointed out that I was not an experienced professional UFO researcher like some of the other witnesses.

"And I have not yet had the privilege of flying a UFO, nor of meeting the crew of one, but I do feel that I am somewhat qualified to discuss them since I have been into the fringes of the vast areas in which they travel.

"Most astronauts are very reluctant to even discuss UFOs due to the great numbers of people who have indiscriminately sold fake stories and forged documents abusing their names and reputations without hesitation. Those few astronauts who have continued to have a participation in the UFO field have had to do so very cautiously. I have always been honest about my views on the subject. It's up to each of us to say what he believes in. There are several of us who do believe in UFOs and who have had occasion to see a UFO on the ground or from an airplane."

After describing my experiences with UFOs in the skies over Europe, I suggested that if the UN agreed to pursue this project and to lend its credibility to it, many more well-qualified people might agree to step forth and provide information and assistance.

When I finished, I was asked by a panel member a question I'd heard a hundred times before: Had any U.S. astronaut ever seen a UFO from space? I told of the one occasion (*Gemini 4*) that might have been a UFO sighting.

More experts testified, but the hearing ended with no recommended course of action by the panel. Sadly, the United Nations never followed through on UFOs.

It may have had more to do with priorities and ongoing world politics than anything else, since there were a number of more "earthly" problems that needed immediate attention. It's possible that UN members felt they didn't have the monetary resources or political mandate to look into the global UFO phenomenon.

Whatever the reasons, UFOs have yet to be studied the way Dr. Hynek so eloquently called for: from an international perspective, with scientists and researchers sharing information globally. It was a dream of his that Dr. Hynek did not live to see fulfilled.

Twenty years later, on June 29, 1998, this *Washington Post* article appeared:

UFO MYSTERIES DESERVE SCIENTIFIC STUDY, PANEL SAYS

Some supposed UFO sightings have been accompanied by unexplained physical evidence that deserves serious scientific study, an international panel of scientists has concluded.

In the first independent scientific review of the controversial topic in almost 30 years, directed by physicist Peter Sturrock of Stanford University, the panel cited cases that included intriguing and inexplicable details, such as burns to witnesses, radar detections of mysterious objects, strange lights appearing repeatedly in the skies over certain locales, aberrations in the workings of automobiles. . . .

I am left to wonder: *What will it take?*

Must a UFO land at the Super Bowl to get the world's undivided attention?

FLYING SAUCERS: MADE IN THE USA

In 1978 I went to Utah to visit a man who had supposedly had a life-altering UFO experience years earlier.

A business acquaintance, Greg Linde, president of Southern Pacific Land Company, a subsidiary of the giant railroad, asked me to come with him to visit Wendell Welling, a businessman who lived about seventy miles north of Salt Lake City. I was told that as a result of Welling's UFO experience, he'd started building his own saucers and claimed to have a workable design.

Given my own experiences chasing saucers, I was not about to discount the story. Also, that a top executive of Southern Pacific considered this guy's story credible enough to ask me to take a trip with him meant something. But no sooner had we scheduled the trip than we received bad news: Wendell Welling, in his sixties and described by his wife as "in the prime of life," had dropped dead of a heart attack.

I was disappointed, but having never met Welling, I didn't think much more of it. A few months later, Linde called and said that Welling's son-in-law wanted us to come to Utah anyway and see the saucer designs. I jumped at the chance.

I flew commercially from Los Angeles, where, since leaving NASA, I had settled with my second wife, Suzan, a schoolteacher. With Suzan, whom I married in 1972, I finally found marital happiness and was blessed with two more daughters, Colleen and Elizabeth. (Trudy and I divorced shortly after my retirement from NASA. A commercial pilot, she stayed on in Houston, where she ran her own Learjet charter service until her death from cancer in 1994.)

At Salt Lake City I was met by the railroad executive, and we flew into rural Tremonton Airport, about twenty miles south of the Idaho border, on Southern Pacific's corporate twin Cessna 310. When we landed we were met by Welling's son-in-law, Scott Holmgren, a well-spoken man in his thirties.

We drove through picturesque farmland. Along the way I learned more about the late Wendell Welling, whom his son-in-law described as "self-taught" and "intensely curious." Welling had been a successful farmer, trucker, and grain dealer, owning farms and grain elevators in Utah and Idaho. In the last years of his life he had developed and opened Belmont Springs, a resort and golf course in northern Utah. But the most monumental year of his life had occurred two decades earlier.

According to his son-in-law and to Welling's own detailed notes, copies of which were provided to me, the day that forever changed his life was October 6, 1959. Welling was standing with two other men, George Nelson and Walter Buhler, at the entrance to an empty grain elevator several miles outside Montpelier, Idaho (population 2,604), when they heard a loud engine start up. They were aware that the cavernous metal structure they were standing next to, when empty, could act as a huge "ear"—catching and amplifying sounds from a good distance away.

The engine noise seemed to be several miles to the northwest, near the south end of Nounan Valley. The three men knew there was nothing back there in the way of homes or farms and began to speculate about what type of machinery with an engine that size could be back in there—and why—when a second engine started up.

These men knew machinery. The engines seemed to be the same size, and by the sound of their exhaust they had to be big. The men could tell that whatever load the engines were beginning to pull was a heavy one, and they were probably pulling it uphill. Soon the engines were emitting the most powerful exhaust cackle any of them had ever heard.

Suddenly the noise of a tremendous air blast came across the valley. It sounded like a high-pressure air tank with several valves that blew out all at once. It lasted for only a second, and then everything went quiet.

The men waited expectantly.

Although the sun had been down for about twenty minutes, it was still light enough to see. The sky was cloudless except for a couple of thunderheads some eight to ten miles to the north.

Welling, looking westward at the eastern slope of a range of steep hills in the direction they believed the noise had come from, saw only shadows. He was studying the terrain, looking for the source of the noise, when his gaze went to the sky above the hills. Hanging motionless above the crest was a large saucer-shaped craft.

He called it to the attention of his companions, who also spotted it.

The shape began to rise slowly, then stopped. Then it started coming their way, losing altitude and building up speed along the way. As it came closer, they saw that the front of the object was tilted down slightly.

It swept eastward across the valley, not leveling out until it was about a thousand feet off the ground. It flew at this altitude past the men, about two city blocks from their position, completely soundless now, and traveling an estimated four hundred miles per hour.

The three men later described it as at least a hundred and

sixty feet in diameter and some twenty-two to twenty-three feet thick from top to bottom. The craft was shaped like two plates put together face-to-face, with the top plate being upside down and slightly wider and flatter than the bottom plate. The plates didn't come quite together, and the top plate was spinning while the bottom plate stood still. In between the plates was a separation of about two feet, with some vertical posts visible. The posts were spinning along with the top plate.

There were no windows of any kind that they could see, but set into the belly of the craft were three disks six feet in diameter; later the men agreed they might have been retractable landing gear of some sort. The craft was a light gray with a tint of blue and had no markings of any kind: no numbers, no writing, no insignia.

As it passed the men, the craft made no sound, not even the slight whirring or disturbance of air one would expect from normal aircraft. There was no vapor trail, smoke, or exhaust residue coming from the craft. It just floated by as though the wind were moving along with it so as to cause no disturbance whatsoever.

The three men stood in utter astonishment, not able to believe their eyes but afraid to look away and lose sight of it. At that point they found it too fantastic for words or expression, and no one said anything. They just watched.

The saucer crossed the valley from west to east, traveling some eighteen miles in two to three minutes. As it neared the hills to the east of the town, it rose up effortlessly and sped away to the southeast, passing from view.

For three days following the sighting, Welling dropped everything else he was doing and feverishly wrote down pages of his impressions, which he described as "things I think we should remember and know about this craft which could lead us to discover just what took place that evening."

Wendell Welling had no background in architecture, engineering, drafting, or aviation. By all accounts he was a doer, not a dreamer, and after that October day in 1959, he became a man searching for answers to questions he'd never known existed. Admitting to being lost in "wonderment and almost disbelief" at what he had seen, he sought information about aerodynamics and aircraft design wherever he could find it, scouring libraries and taking home stacks of books to study at night.

"He did not take his experience lightly," said his son-in-law, who had worked closely with Welling on the saucer designs. "He thought he saw something important that night, and he was determined to re-create what he'd seen."

"I began to come up with possibilities of how this craft we saw might have accomplished this feat," Welling would later write. "Each new concept brought up other obstacles which for a time seemed to rule out any further search along these lines, when suddenly a way to jump this hurdle would be seen. But when I felt that I had cleared all the engineering obstacles, I still didn't have faith in my reports and drawings. How could this type of craft possibly lift anything, when it couldn't throw the air straight down, but had to throw it off almost on a horizontal plane?"

But Welling found that he couldn't quit.

"I had seen it with my own eyes, and three of us together watched and listened to it as it took off. Then it flew right past us. Once you know something, you sure find it hard to forget, especially when it entails a ship that only Buck Rogers could dream of."

And so Wendell Welling began to build saucers. He hoped they might furnish some answers to whether "a craft shaped like the saucer I had seen could show lift on a test stand, and whether the lift potential could be ascertained."

We arrived at Welling's six-hundred-acre wheat farm, situated in a fertile valley with mountains on either side. Inside a large, barn-like building, I came face-to-face with a collection of genuine made-in-the-USA flying saucers.

Some of the early ones were small, no more than two feet

in diameter. The largest disk, fifty to sixty feet in diameter, had been under construction when Welling died. There were several completed models of varying sizes up to twelve feet across.

I reviewed schematic drawings Welling had made to figure out how the size and weight of the vehicle affected lift, how much horsepower was required to drive it, and the revolutions per minute (rpm) needed to achieve lift and increase the vehicle's "tip speed"—the speed out on the periphery, or spinning lip, of the saucer.

Welling had found that using a dome-shaped top and relatively flat bottom caused a saucer-shaped craft to act as a round one-piece airfoil—something like a Frisbee. Comparing it to a vertical wheel, which had proved to be an effective means of moving things on the ground for thousands of years, Welling had concluded that a "horizontal wheel" was an efficient airfoil with which to travel through the atmosphere.

I saw his point, and it seemed so logical as to be obvious. The lift area of a winged aircraft was limited to the underside of the wing and tail section. A saucer of equal "wing span" had much more lift area: 100 percent of its surface provided lift.

Welling had incorporated into his design the "spinning top" he had seen on the UFO, putting a lot of thought into what it did and why it was important. His theory was that the spinning platter on top took advantage of a buildup of kinetic energy, equating it to what happened with a favorite children's toy: the spinning top.

He was certain that the giant "fly wheel" on the saucer he had seen furnished the thrust that lifted and drove the ship. He believed that the engines he and the other witnesses had initially heard were warming up the saucer's huge spinning top—he estimated it might have weighed as much as twenty tons—to the necessary rpm before shutting down. "Once that weight is spinning like a huge top," he wrote, "the energy requirement to hold the rpm drops way down and the craft is able to take advantage of a great power potential with the fly wheel acting as a centrifugal pump, battery, and huge gyroscope all wrapped into one."

In theory, I found Welling's descriptions sound. He had spent the last years of his life and thousands of dollars of his own money to prove these theories, and I now wondered: *Did they fly?*

I was invited to take the controls of the largest "flyable" saucer in the barn and give it a whirl. Approximately twelve feet in diameter, the saucer was constructed of silk cloth stretched over balsam wood, making it very light. The bigger saucer still under construction was being crafted of more durable materials.

Welling had not attempted to develop a propulsion system. Having focused his efforts on aerodynamics, he chose to power his manmade saucer by an electrical generator that drove an onboard fan, which pushed cold air around for thrust.

In addition to a heavy-duty electrical cord that ran from the generator to the fan unit, the saucer was tethered to the ground by a half-inch steel cable that looked to be about twenty feet long— meaning that I wouldn't be going for any altitude records.

I sat at the control station about ten feet away. The sole "flight control" was an airplane-type stick. Pulling it back increased the rpm to the fan and provided greater lift, I was told; pushing it forward decreased lift.

When the generator was fired up, the top section of the saucer began to spin, just as in the big saucer Welling had reported seeing twenty years earlier. Then the bottom section started spinning too, so there were two counter-rotating disks. In Welling's design, the speed with which the bottom disk spun controlled the slots that the air flowed through. The cold air flow, which kept it airborne, was pushed horizontally over the curved surfaces of the facing plates and downward toward the ground. The more air that was moved, the greater the lift.

The only noise in the room was the slight whir of the generator.

I applied gentle backward pressure on the stick, and the saucer jumped off the test stand, soundlessly, and rose into the air effortlessly to ten feet or so. I was amazed at the ease with which the bird took flight. Up and down it went as I moved the stick forward and back.

Compared with other aircraft I'd tested, this was a pretty crude model. But from the moment I first pushed back on the stick, I was extremely impressed with the saucer's lift capabilities. With very little power—the fan wasn't even powerful enough to move air effectively through a large room on a hot day—this thing flat-out *flew*.

There was a device—a counter-balanced scale with weights of known quantity on the other end—hooked up to the saucer to measure the lift it attained in pounds, but I didn't need to check the numbers. I could tell by the feel of the stick. I flew the saucer for about ten minutes, and the experience really opened my eyes to what a vehicle of this configuration would do; specifically, the tremendous lift that could be developed from the saucer shape.

Boy, I thought, *we've been going the wrong way all these years with winged aircraft*.

Before we left that day, the son-in-law told us that there had been a multimillion-dollar offer from a Middle Eastern country to buy all of Welling's saucer designs and prototypes. He said he knew his father-in-law would have preferred keeping the technology in this country, and wondered if there was anything we could do to facilitate that. I couldn't think of a way to help him. It wasn't something I could take to NASA. As I said, these were pretty crude models, and money would have to be spent to develop Welling's idea further. As for Greg Linde getting his company interested—"I handle land holdings for a *railroad*. My board of directors would have my hide if I spent money on flying saucers!"

For years I wondered if that sale ever went through—if in a

faraway desert, an oil-rich country had figured out a propulsion system and was flying advanced versions of Wendell Welling's saucers around that part of the world or beyond.

I only recently learned that the sale did not go through. In fact, I understand that my "test flight" was the first and only one of a Welling saucer after his death. According to his widow, Elsie, all those "round things are still around here somewhere."

Wendell Welling had risked a lot, not only in going public with his UFO experience and taking the chance of being labeled a crackpot but in expending his time and money and placing his reputation on the line by building those "round things." But build them he did, for his neighbors and the world—even a former astronaut—to see.

Whenever I am asked, as I often am, if I believe that some saucer-shaped UFOs might actually be highly classified U.S. experimental aircraft, my answer is always the same: "I sure hope so." Considering all the taxpayers' dollars we're spending, I would *hope* somebody is doing something worthwhile. I hope we're building our own saucers. Who is to say that the big saucer Welling had seen wasn't such an experimental craft?

Ever since that day in Utah, I have been convinced that saucers are the aircraft design of the future—both for this world and for travel beyond.

As for Welling, the last paragraph of his notes was poignant: "I might make mention here that to try to make someone else believe this [UFO sighting] is a tremendous feat in itself. We have discovered that over a telephone or by letter it is impossible. In fact, it embarrasses them and yourself, because you can tell that the party on the other end of the line is wondering what kind of psychopath is on the other end."

Clearly, the grain dealer had become a believer. I was sorry I didn't get to Utah in time to meet Wendell Welling. I'd like to have been able to shake the man's hand and tell him I believed him—and that he'd made some damn fine flying saucers.

HELP FROM THE COSMOS?

The middle-aged man could have been waiting at any bus stop in America that summer day. He wore chinos, a faded shirt, and tattered gym shoes. His Coke-bottle glasses had been repaired at the temple with black electrician's tape. He walked into my office at the Disney Imagineering complex in Glendale clutching a brown paper lunch bag as if his life depended on it.

I had gone to work for Disney in the mid-1970s as vice president for research and development. The word had gotten out the past couple of years that I had an open-door policy when it came to considering new technologies from any and all sources.

After brief introductions, the guy explained that he'd been working for years on an "advanced engine" but had found no one willing to give him the time of day.

I had heard much the same from others who had walked in off the street with some of the most fantastic things I'd ever seen. Without college degrees, financial backing, or high-powered contacts, many of these folks had found it difficult to be taken seriously. It was a shame. In the history of the world, lots of "little people" had been responsible for some big breakthroughs. How could any of us be too busy or too important to be receptive to new ideas for improved technologies?

I listened, wondering if he had his "super engine" out in the parking lot. About then, my visitor reached into his lunch bag and pulled out a tiny reciprocating engine. It started right up.

I couldn't believe my eyes. It was the smallest working four-cycle motor I'd ever seen. But work it did—humming along like the Little Engine That Could. The backward and forward motion of its tiny pistons produced rotary movement of a crankshaft about the width of a pencil.

I showed my amazement, and the inventor swelled with pride.

Disney was building lots of new rides and exhibits at its amusement parks, I explained. Taking down his name, phone number, and other pertinent information, I promised to call in the event that we needed self-propulsion on a small scale.

I meant it too.

Before he left, the man shut down his engine and returned it to the paper bag. He walked out with a new bounce to his step. He and his invention had been appreciated, if not bought, and he was pleased to have been recognized at last.

For me, it was still another reminder to avoid making a judgment based on someone's appearance. Who knew what brilliance lurked behind the patched-up horn-rim glasses of the guy waiting at the next bus stop?

Another amateur inventor had come in with a stereo-optic camera and projector that allowed viewers to see a movie in 3-D without wearing special glasses. He accomplished the effect with two lenses in the camera and projector and by projecting the motion-picture images onto a special rough-textured screen. This idea we bought, and spent money developing it. In the end it went into a warehouse and never saw the light of day. I believed the decision was a political one—the capital investment that would be required of the theater owners, who are a powerful industry force, was probably what killed this new, potentially revolutionary movie system.

Nevertheless, big things were happening at Disney at the time. We had a large R&D budget and were working on a number of alternative energy projects, including an electric vehicle, solar-powered systems of different kinds, and experimental residential use of my old friend from *Gemini 5*: the fuel cell.

We had taken a small tract of Florida homes off the public utilities grid and wired each one with its own individual fuel cell. The fuel cells supplied all the electricity for the homes, and because they ran on natural gas, there were no residual pollutants at all. I was perplexed about why it had taken all this time for the fuel cell, which had made trips to the Moon and other extended manned space exploration possible, to be put to such a simple commercial test. We found them to be very proficient, although the particular brand we were using turned out to require quite a lot of maintenance and repair. Fuel cell design has since improved greatly, and I expect to see them in homes of the future as well as in electric cars, thereby eliminating the need for banks of batteries that need to be recharged every hundred miles or so.

Epcot Center was coming together, and lots of interesting projects were flowing from that development, which involved alliances with major companies such as General Motors, General Electric, and RCA. One afternoon several of us from various companies and in different locations held the world's first closed-circuit satellite conference in real time. I had access to the inner sanctums at Disney and other firms, and felt I was involved in some interesting work.

I was still flying whenever possible, and Disney had its own plane, which was available to me from time to time. But I did miss NASA's T–38s—and having my own jet to run around in. Once you've flown a jet fighter and gone to afterburner just for kicks, no other airplane matches up. A fighter is not unlike what they say about a stunningly beautiful and mysterious woman: exciting and at times unpredictable, but worth the danger.

For several weeks in 1978 a woman whose name I didn't recognize had been trying to reach me at my office, but I was always tied up or away and she declined to leave her number. One morning she called and my secretary put her through.

The woman introduced herself and invited me to lunch.

"*Lunch?*" I said, somewhat taken aback. "What's this about?"

"I think we have some mutual interests."

As I said, there's something about a mysterious woman. And Valerie Ransone turned out to be beautiful too. She was a brown-eyed blonde in her late twenties who bore a resemblance to the French actress Catherine Deneuve.

We went to a barbecue place for lunch. It didn't take long for me to appreciate that this was one smart lady. With a master's degree in broadcast journalism, she had worked for network radio, covering a Midwest news beat during the Watergate scandal. She then worked for the White House during the Ford administration, helping develop a national energy conservation program, with an office in the new Executive Office Building next to the White House. As a strategic planner, she worked on devising ways to better educate Americans about energy concerns and undertook a thorough examination of big oil and the role of Shell, Exxon, and Mobil in the worsening energy crisis that would lead to long lines at gas stations. Valerie said she'd investigated alternative energy sources and technologies, like solar collectors and wind generators, but found most of them too small-scale to service the demand.

Valerie had held some important positions for her age, although she carried herself not with arrogance but with the air of someone who knew her place in the world. She also wasted little time getting to the point.

"I have a plan to bring together a group of technical people who have unusual talents," she said before our salads even arrived. "I know Disney is doing some interesting things, but there are other technologies vastly more advanced. I'd like you to consider helping me assemble the nucleus for a private technology group."

"I already have a job," I said.

"I'm talking about *advanced* technologies. Technologies we need to understand if we are to solve the problems that threaten mankind's potential as a race."

"Uh-huh."

Her eyes bore into mine.

"You must understand, Colonel. I have access to unusual help."

I nodded.

Then she gave me both barrels. "My source of knowledge is not of this planet."

She knows my feelings about UFOs and is putting me on. It was my turn to look hard at her. She didn't flinch from my icy stare.

From meeting the stream of people off the street with their inventions, I knew that out of every ten people, three or four were various shades of nut cases. Another three or four—like a fellow who had come in that morning with batteries vastly superior to those made by Ever Ready—would be for real. I didn't yet know to which group Valerie Ransone belonged.

"Sounds interesting," I said as casually as I'd say, "Pass the butter."

"There is a universal intelligence that permeates the Earth. The source can originate from any one of numerous points. I know because I am getting these signals and have been for years. I think we're involved in a grand communication experiment." With a smile, she added, "This experiment moves the idea of broadcast journalism to a whole new level of possibility."

Who is this articulate, bright-eyed woman? I wondered.

"The point is," she went on after the waiter came with our

entrees and left, "there is a significant intelligence source in the universe that wants us to succeed. They're willing to serve as intermediary so we can evolve as a people and a civilization, but we need to do the spadework for ourselves. No one is coming here from another planet to do it for us. And time is running out."

She said a parent group had been formed with offices in Washington, D.C., where she had an office. She explained that a subsidiary, the Advanced Technology Group, would be responsible for testing and implementing new and revolutionary technologies.

"We're putting together a network of thinkers, innovators, engineers, teachers, scholars, scientists, and social policy analysts who are ready to contribute their research to help design a new landscape for tomorrow. We're going to bring together talented people and concepts for high technologies that are very exciting. Everyone will have their own role—their own area of expertise. Propulsion. Medicine. Electronics. And most important: energy technology—to provide us as a nation, and as a globe, with a clean, renewable, unlimited source of power. We can use your help."

"Why me?" I asked.

"Your credibility from the space program would help attract the best technical people possible," she said. "They'll need to build new equipment to prove some of these theories to the world. We'll provide the context for that to happen."

I dove into my barbecued beef thinking it might be wise to get lunch over with, but then a funny thing happened. The more she talked, the more intrigued I became. I had enough skepticism that I was simply listening, nothing more. At the same time, my lifelong openness to new and unusual possibilities—both on Earth and in the vastness of space—kept me in my seat.

By the time we finished lunch, she'd piqued my interest sufficiently that I agreed to meet with her again, and to keep listening.

We met several times over the next few months. Each time we parted, I was even more impressed with her intelligence, insight, and enthusiasm. I saw no weirdness, foolishness, or quackery. Valerie Ransone was a thinker and a doer—a nice mix.

Along the way, she introduced me to a number of individuals who were willing to contribute their expertise to her effort. I found them to be technically very qualified: outstanding scientists and researchers, many with private, university, and even military lab affiliations. In some cases they had been ostracized by their peer groups and professional organizations because they were "way out there" in their theories. The way-out part didn't bother me as long as their math and science were sound. That they appeared to be well qualified, and at the same time believed in Valerie Ransone and were ready and willing to join the Advanced Technology Group, meant something. Some had known her for years, and were true believers. I saw what she offered them: hope that they might be able to prove their own and other new technologies to make the world a better place.

They were ready to take the leap.

I was not yet there; I needed more information. It came slowly, as in working a giant jigsaw puzzle one piece at a time.

Valerie Jean Ransone was born in Illinois to an upper-middle-class family. Her father was an engineer and her mother a stay-at-home Mom. Valerie had always been a quick learner and excelled in school. I learned that she was born with some unique abilities, such as a photographic memory, a strong sensitivity to "energy fields," and an "openness" to assorted "electromagnetic signals." She explained matter-of-factly: "Some of us just have bigger antennae."

She confided in me that when she was seventeen years old she'd had her first "contact experience" while driving home alone after attending a summer party. She told of experiencing six hours of "missing time," and said that her life had been irrevocable changed.

After we had known each other a few months, Valerie went into more detail about her "first contact." She said a "space civilization" of beings more advanced than humans had contacted her. This civilization represented the "highest minds in the universe" and was seeking individuals with unique capabilities: people with whom it could communicate via telepathic transfers, providing technological and other information. She referred to her "contact group" as UIC—Universal Intelligence Consortium. "This intelligence consortium presented this to me as a matter of world peace. I was told that if we succeeded it would be a natural progression of man's evolutionary processes. If we failed, the possibilities were unthinkable. I felt I had no choice but to stumble along and pursue this course in good faith."

Valerie said she agreed to be used as a "telepathic conduit." She later found out, she explained, "that I wasn't the only one who agreed to participate in the experiment. The communication plan, as it was outlined to me back in 1968, sounded rational. As rational as anything could sound to a seventeen-year-old."

More than once she repeated to me her original pledge, made upon that first contact, that she would participate in the "communication experiment" as long as no one was harmed and only if the process contributed to world peace. "I was attracted by the idea of world peace amidst the ugly backdrop of the Vietnam War," she went on. "Here was something I could do, though I must admit that had I known what I know now, I would have said I wasn't up to the task. I've never had a problem receiving signals, although it can be physically exhausting. The problem comes when the consciousness returns to daily living—the inordinately difficult task of clearing my mind of everything I've taken in and then carrying on a normal conversation with people or just going to the supermarket and buying broccoli or taking the cat to the vet. Often I would receive contradictory information, telepathically, from what was emerging from a person's mouth who was standing directly in front of me.

The benefit was that my BS meter worked all the time, seven days a week, without an off switch."

Her establishment credentials disarmed many, Valerie said with a mischievous smile. "Because of where I came from, I was allowed entry into some former male bastions of power that had been unaccessible to women."

Brains and beauty are a powerful combination, I thought, while considering the astonishing scenario of a well-educated young woman working at an important post in the White House while claiming to be in contact with extraterrestrials.

It made Watergate seem awfully puny.

After leaving the White House, Valerie taught communications and media production at a private college in Washington, D.C. One day, following her lecture about dolphins and inter-species communication, a student told her of a lecture that evening by a well-known medical doctor and researcher with a long-time interest in paranormal communication. Valerie attended the lecture that night by Dr. Andrija Puharich, a graduate of the Northwestern School of Medicine.

Dr. Puharich, who had a long-time interest in studying psychic phenomena, told of working with some gifted "space kids"—"all with big antennae of their own," Valerie explained—brought in from seven countries to take part in extended scientific experiments in Ossining, New York, not far from New York City.

After the lecture, Valerie went up to the podium and told Puharich, "Doctor, I'm one of your space kids." He invited her to visit his lab, and soon came to value her telepathic abilities as well as her administrative skills. He hired her as a research assistant in 1977. Her job was to identify, document, and test many of the space kids who came for regular evaluation to the private lab at Puharich's estate compound.

I learned that Dr. Puharich was for real and was doing research on contract for the U.S. government—even the military. Valerie arranged for us to meet in Washington, and I found him to be brilliant. We discussed electromagnetic energy and talked for an hour about various advanced propulsion systems for space travel.

Dr. Puharich had previously discovered a young Israeli who had gone on to fame: Uri Geller. Like Geller, Valerie was able to bend utensils with mind power.

"I wasn't too good with spoons," she told me at one of our lunches, laughing. "I specialize in forks."

I wanted to know how it worked.

"Anyone can learn to do it. It's just mind over matter."

"Show me."

Rubbing her index and middle fingers over the hump part of the fork but not touching the metal, she closed her eyes, and within thirty seconds the fork began to bend and kept moving until the prongs nearly touched the handle.

This was the restaurant's fork, not Valerie's, and she had accomplished the feat without touching the utensil. I was convinced I had seen the real thing, not a parlor trick.

Valerie described the space kids as "long-time tuned in." They spent time in Ossining perfecting their telepathic skills. "Telepathic powers," Valerie went on, "are part of human evolution. The potential lies in every human being. You just need to eliminate the distraction." They also practiced "remote viewing," in which the power of the subconscious is used to "travel" to different times and places and "see" actual events. (The U.S. military and CIA had secretly become involved in remote viewing research and, unbeknownst to Valerie at the time, was funneling money into Puharich's private research group in the furtherance of what became known as "psychic espionage." The CIA's program, known as Stargate, would not become public until fifteen years later, and only when former government "remote viewers" began to come forth with their stories in articles and books.)

Puharich conducted experiments in a "Faraday cage,"

named after Michael Faraday, who in 1831 rocked the scientific world with his discovery that magnetism could produce electricity if it was accompanied by motion. The cage was a rectangular metal box approximately eight by eight by twelve feet, which was lined with copper and placed on insulated supports. Inside was a complete electrical vacuum—no electromagnetic waves such as TV and radio signals could penetrate the cage. This was the environment for various communications experiments to see if the subject was picking up signals from other sources of intelligence.

"With the shielding, communications came through much clearer when we were in the cage," Valerie said. This was important, she explained, because it served as a "double-blind experiment" to isolate signals, making certain the subject was not "reading information from the biofield" or local environment. "Psychics can read off someone else's field. We didn't want that. We weeded out the fakes."

Valerie explained that "space civilizations of beings more advanced than we on Earth" had made contact through telepathic means with members of the research group. During her time in Ossining, she said she documented twelve civilizations that appeared to have "receivers or agents of change" operating on Earth, "a sort of relay network." These receivers were particularly clear, she said, among the youth—the younger space kids who had sought out Dr. Puharich to have their extraordinary, and at times disturbing, natural gifts validated.

Valerie said it had been a priority of hers to document the messages that "came through the space kids," making sure that no one—not Dr. Puharich or anyone—was planting subliminal or hypnotic suggestions in the minds of these subjects. "I became one hundred percent certain that no suggestions were being planted," she said.

In all, she worked with some thirty-five of Puharich's space kids, ranging from twelve to fifty-five years of age. She helped collect the information that came through each transmission and cross-correlated the data with the content of other transmissions to isolate the "civilizations with whom communication was occurring."

Using her "own access to extraterrestrial sources," Valerie would attempt to confirm or deny the authenticity of information contained in Dr. Puharich's Faraday experiments. Often this information would be of a highly technical nature. Without formal training in science, she would use other resources and experts in various disciplines to validate the accuracy of the content—for the most part, she explained, "backing into the content" because she didn't know what it meant until it was analyzed by the experts. Often, she said, the answers and questions that "came through" were offered by "other intelligences" without any formal inquiry having been made. "I began to wonder if our minds weren't being read," she told me.

Valerie explained that she had for some time been "pretty much a computer," bringing in transmissions in "several different languages." She didn't use the term *channel* to describe what she did, and she attached no mysticism to it. She preferred to be seen as a "pioneer in a new field of communication: *interdimensional communication.*"

And suppose it was true?

I believed that the big question everyone liked to ask about when we might make *contact* with an extraterrestrial intelligence wasn't all that weighty. There had been too many credible people with proven cases of sightings and contacts with UFOs, flying saucers, and other vehicles clearly not of this world to keep debating that issue. There has been *contact*—period. I always felt that these were the more important questions: *Now what were we going to do about it? Was there a way we could sit down with them and learn about some advanced technologies?*

Suppose everything Valerie Ransone said was for real?

What if she and some other select people *were* receiving

signals from an advanced civilization?—the kind of alien signals NASA had been trying at great expense to pick up for years. Suppose they were just coming in on different wavelengths— not radio frequencies that could be picked up by SETI's radio astronomers but via telepathic messages to people who had such communication skills?

What might that kind of assistance mean in the history of mankind?

The heck with getting to the Moon first, or even a Mars landing. This would be bigger. *Much bigger*.

I had been talking about UFOs and the very real possibility of extraterrestrial intelligence for years. Was it time for me to put up or shut up?

I had long believed it vital to keep an open mind and not lock up on the status quo. Remarkable new technologies that we can't envision in the present are always around the corner. I had seen that in the space program. Without open minds, we'd still be gazing up at the Moon wondering what kind of cheese it was made of.

What interested me most was the promise of new technologies. Our track record in this country is dismal when it comes to developing and utilizing advanced technologies. Look how agonizingly long it took the fuel cell to make its way from space to everyday life, even in a small way. Throughout my military career and years with NASA, I saw people willing to stifle advanced technologies in favor of business as usual. I had fought such narrow-mindedness with every ounce of my being. When I was a test pilot at Edwards, I worked on the early digital electronic flight controls and found them much more efficient, and a lot safer, than the traditional hydraulic controls. It was an improved design change that should have been instituted in new aircraft *immediately*—for safety alone. But how long did it take? *Thirty-five years*. Seeing such vastly superior technology ignored by the "experts" had been a nearly lifelong frustration.

Valerie was clear about what she hoped to accomplish. She envisioned the group building prototypes of electromagnetic propulsion systems and other free-energy devices; developing technologies to conserve Earth's natural resources; conducting contract research and development work with governments and scientific institutions around the world. The possibility of using medical devices with low-frequency pulses to heal wounds and stimulate bone growth particularly interested her. She also favored advanced research on paranormal phenomena and establishing a clearinghouse for information on new technologies to be shared among private firms and government agencies. "I have been led to believe that our planet and inhabitants can be healed, physically and mentally, and evaluated spiritually so that open communication with extraterrestrial civilizations can be realized within our lifetime. Whatever small part I have in making that happen—well, that's my goal, Gordon."

For my part, from observing Valerie and the technical contributors in their animated discussions about theory and science, I realized that exciting possibilities could be on the horizon. She had already proved to many of these experts an ability to produce vital bits of information that bridged gaps in their existing research. Beyond the question of where this data came from, the fact was that these researchers felt they could take the information and develop some extraordinary new technology.

I was willing to at least give that part a try.

When I asked where the money was going to come from to finance research and development and pay salaries and expenses, Valerie shrugged.

"You've got me. What I've been saying all along is that when the right people are gathered, the money will be there."

I thought that was being extremely hopeful, if not naive. I assumed that short of an extraterrestrial civilization making greenbacks or gold bars for us, we would at some point have to solicit financial backing. But what would it cost me except some time and effort? While I was in a full-time job at Disney, I had flexibility in my hours—and there were always weekends. As long as I could work things around my schedule—

"We'll work around your schedule," she promised.

I agreed to come aboard, ready for what, exactly, I did not know.

THE SPACE SHUTTLE
TRANSMISSION

"There could be trouble with the space shuttle."

It was December 1978.

By then Valerie Ransone and I had been working for a solid year and with great effort to get the Advanced Technology Group up and running. Typical of a new business in the planning, some things had gone smoothly and others had not. We had a handpicked team of top-caliber researchers and inventors who had agreed to come aboard. We had a targeted start-up date and had secured some early financing.

"What kind of trouble?" I asked her over the phone.

Valerie was at her office in Washington; I was at mine in California.

"Technical flaws," she said. "Something to do with the heating or cooling system. It's pretty sketchy."

The warning had come to her during one of her "transmissions." She had no idea when they would arrive—the telepathic messages she believed were from an extraterrestrial source of intelligence. They sometimes reached her at inconvenient times—

such as while driving or in the shower—when she had no way of writing down the often technical and detailed information. She made a point always to document the details as soon as possible, usually typing up the complete messages.

She had shown me some of these transmissions; others she had read over the phone. At first I didn't know what to make of them. Some were very involved, giving advice and specifics about people and events to the extent of mapping out various business strategies for the Advanced Technology Group and even approving or disapproving our list of possible contributors.

Yes, fine man.

Not at this time.

Boisterous, overbearing, inventive . . . OK investment source.

(And yes, I *had* been approved.)

In one transmission Valerie learned that one of our key contributors, Al Jacobs, was facing a grave health threat—specifically, an undiagnosed brain tumor. Not feeling well, he had been in for various tests, but doctors hadn't found anything. Valerie's information was that he should have an immediate CAT scan and that the tumor would be found behind his right ear. She wasn't sure how to proceed—should she call Al's wife or try to speak to his doctor? She did, in fact, attempt to deliver her message, but to no avail. Al eventually had surgery, and a malignant tumor was removed from the right side of his brain. He went blind and died soon after.

Such experiences with her transmissions were sobering.

Four months after mentioning the vague possibility of a problem with the space shuttle, Valerie was in Los Angeles and came by my office. She was worried about new and more detailed information she had received. She showed me a single typewritten page of notes she'd made after receiving the transmission.

This time the warning was quite specific.

In terms worthy of any graduate engineering class, "serious technical faults" were outlined in detail, specifying what could happen during reentry to the system that provided cooling to the cabin and sensitive electronics. The source of the problem seemed to be the electromagnetic effects during space flight on the iron rods used in the cooling system.

At the top of the page was a detailed drawing of a tube of some type, showing its iron-rod center and a chamber that held a liquid substance.

"Who drew this?" I asked.

"I did," Valerie said. "What is it?"

"I don't know."

"What does this all mean, Gordon?"

I didn't have the faintest idea; neither, apparently, did she.

The key was the ventilation system, according to the message. If it wasn't at the proper temperature during reentry, the result could be a toxic release that would fill the lungs of the crew and quickly render them unconscious. No doubt was left that an urgent design change in the space shuttle was called for before another mission.

The transmission warned of the danger that premature launches of experimental components would set up in-flight difficulties that could not be handled by the crew, the result being a catastrophic event that would lose public support for the program. The source of this information professed to being concerned that the future of manned space travel not be jeopardized by such events.

The explicit warning made me think of *Apollo 1*, and what might have been done, even at the last moment, to avert that disaster by taking more time to sort out the long list of problems that had been identified.

I'd been around Valerie long enough to find her knowledgeable and trustworthy, and I wasn't about to discount her transmissions. In addition, her telepathic powers had been proved time and again. She'd say that something was happening at that moment on the other side of the country—at a meeting, for example—and when I checked it out with the participants, even those who didn't know Valerie and had no way of being in touch with her, they would confirm her version of events. I felt certain she was getting good information from *somewhere*; from where and whom I couldn't say for sure.

And now this: a possible catastrophic design flaw in the space shuttle.

For me, this was a moment of truth.

Dare I ignore the transmission, origin unknown? On the other hand, did I march into NASA with the information in hand? Would they consider me some kind of fool—retired from NASA for nearly a decade, coming in with detailed technical information about a spacecraft I had never even flown?

Looking at the intricate drawing of the coil, I knew I had no choice.

I guess it came down to the pilot in me.

I had lost systems at critical times. It could and sometimes did happen, even though everything was designed so that it wouldn't. I knew what it felt like when it happened in space. A systems failure was any pilot's worst nightmare. One such failure could lead to another, then another. When things piled up too much, I don't care how experienced the pilot was, the situation could get out of control real fast.

It's a hard lesson for pilots, but an important one. Our fate is not in our hands alone. We must believe each time we go up that we are capable of controlling all the variables thrown our way. But sometimes it's just not possible. As the pilot, you can be doing everything right and still lose the aircraft—and your life. I had seen it happen to buddies of mine.

You look at every aviation accident and wonder: *If I'd been in that situation, is there something I could have done?* Sometimes you think: *Yes, I would have handled it differently and made it back safely*. But other times you have to admit that placed in the same exact situation, you would have been toast too.

Had I been one of the astronauts aboard *Apollo 1* when it caught fire on the pad, I would have died that day. Saying that doesn't make me feel real good, but denying it is unrealistic and dangerous. In aviation, there are constant reminders that you're not immortal.

Placing the space shuttle transmission into my well-worn briefcase, I flew to Houston to see Bennett "Ben" James, an experienced engineer and supervisor in NASA's Flight Operations whom I knew from Mercury and trusted like a wing-man. Ben was a big guy, well over six feet and built like a Penn State fullback. Unlike some specialists, he was a technical guy who was also good with people.

We sat alone in his office, and I told him the "whole six yards," as we used to say in Oklahoma, where fertilizer trucks carried six cubic yards of material.

Ben already knew about the Advanced Technology Group and our hopes for new technological breakthroughs. In fact, I had been talking to him about joining us after he retired from NASA in a year or so. He knew the name Valerie Ransone and some of her impressive background, but not much more. I now told him it was possible that my business partner was in contact with "higher powers somewhere who may have better information than we do."

Ben, a trooper in every sense of the word, didn't flinch.

"The bottom line for me is who cares where this comes from?" I said. "If it's valid—if it's accurate or the scenario is possible in any way—well, maybe someone should do some double-checking just to make sure things are all right."

"Say no more, Gordo. I agree."

One thing about NASA: they were so accustomed to people doing endless what-if scenarios that no one would sit around wondering where we came up with the possibility of a design flaw on the space shuttle. They would be much more concerned with figuring out if it was a genuine problem, and if so, fixing it. It had been that kind of attitude that had gotten us to the Moon.

I volunteered to help Ben brief several NASA managers. "But I'm not sure we should tell them the source," I smiled.

"I agree."

NASA engineers immediately went to work examining the space shuttle's cooling system, looking at the detailed scenario I laid out for them. They quickly identified and, within days, fixed the potential problem with the cooling system—just as outlined in the transmission I carried in my briefcase.

I was relieved that they found and fixed the problem on the space shuttle. Was I surprised that the cooling system flaw existed? Not really. With Valerie Ransone, I had moved beyond surprise. But the experience gave me another shot of confidence that the source we were getting technological assistance from was for real.

When I told her NASA had located the problem and made the fix, she said simply, "Of course they did." She never raised the subject again.

Valerie Ransone was not an engineer, had never worked for NASA, and wasn't involved in aviation. If this vital and very detailed information hadn't come from a source of higher intelligence that for some reason was monitoring the U.S. space program, then *where did it come from*?

TESLA: TWENTIETH-CENTURY GENIUS

Nikola Tesla was born before his time. He dreamed of inter-planetary communications and space travel even as he battled with Thomas Edison over which method of electrical current would be most efficient for modern civilization.

Time and again, Tesla was proven correct. But history has a short memory when it comes to the Croatian-born inventor—the father of modern-day electrical engineering—whose alternating current (AC) electricity replaced Edison's direct current (DC) soon after the turn of the century.

The advantage of AC power—invented by Tesla in 1882—is that high voltage can be sent hundreds of miles through reasonably sized wires, then reduced for household use by transformers. If the wires accidentally come together, they short out just at the place where they touch and only for as long as they are in contact. In contrast, DC power needed huge cables and power stations every few blocks. Also, the thick cables heated up, and when shorted they melted all the way back to the powerhouse, meaning that streets had to be dug up and new cables laid.

With the stakes so high, Edison and his company, General Electric, put together a traveling road show to demonstrate the "dangers" of AC power, going so far as to publicly electrocute puppies and larger animals—in one case an elephant.

Tesla, with the help of his friend and benefactor, industrialist George Westinghouse, won the "battle of currents" by proving the safety and efficiency of his method when he illuminated and powered the New York World's Fair of 1899 and a year or so later harnessed Niagara Falls by converting its hydraulic power to AC electricity. By 1905 all generating stations in the United States were operating on AC power. Although history credits Edison with inventing our worldwide system of electricity, Tesla's AC power is what runs our cities and households today.

Early in our discussions, Valerie Ransone brought up Tesla's name. Anyone seriously involved in technology has heard of Nikola Tesla, the greatest inventor the world has ever forgotten, and I remembered him well from my early days studying engineering and electrical theory. Through my association with the Advanced Technology Group, I came to have an even greater respect for his work. Valerie was a champion of Tesla's technology, as were a number of our top contributors.

When she was working for the White House on alternative energy, she studied Tesla's plan for a "wireless power system" as an alternative to fossil fuel energy, which was finite and would one day be used up. She'd ended up doing more than a thousand hours of hard research on Tesla, sorting through old files in the Smithsonian, examining his original patents (more than five hundred of them), and running down endless leads to inventors who claimed to have modern-day free-energy devices.

"Are you familiar with Tesla's wireless work?" she asked me.

I said I knew about his accomplishments in radio.

"That's only part of it."

One of Tesla's most revolutionary discoveries was his system for transmitting energy via wireless antennas. In 1900 he obtained two patents on the transmission of wireless energy covering both methods and apparatus and involving the use of tuned circuits as receivers. Two years earlier Tesla described the transmission of not only the human voice—this three years before the wireless radio was "invented" by Marconi, a former assistant of Tesla's— but images as well. Tesla later designed and patented devices that evolved into the power supplies that operate our present-day TV picture tubes. In 1900 Tesla announced that "communication without wires to any point of the globe is practicable."

Tesla's turn-of-the-century experiments revealed that the air at its ordinary pressure is distinctly conducting. In some of his research notes, copies of which Valerie showed me, Tesla wrote, "This opens up the wonderful prospect of transmitting large amounts of electrical energy for industrial purposes to great dis-tances without wires. Its practical consummation would mean that energy would be available for the uses of man at any point on the globe."

The possibilities were astounding—no unsightly utility poles and wires passing over every street in America and no hardwired grid that could fail without notice, stranding a major metropolitan area or even several states without power, as has already begun to happen in the United States.

I learned that Tesla had received early financial backing for a wireless transmission project from Wall Street financier J. P. Morgan. Construction was begun on a full-sized broadcasting sta-tion and 180-foot tower on Long Island, which would have been able to provide usable amounts of electrical power at the receiving circuits. But financial support was suddenly withdrawn by Morgan when it became apparent that such a worldwide power project couldn't be metered and charged for, as could a hardwired grid. The project was never completed, and the Tesla tower was torn down, without question stifling the development of clean energy technology in this country and beyond.

"I want to build a Tesla tower," Valerie said.

Many of the experiments she had been involved in with Dr. Puharich in Ossining were based on the work of Tesla, she explained. Several of the space kids she worked with were "bringing through" Tesla-related information on technology that had "global and interstellar application," she told me. Her interest had been piqued when she found math equations in Tesla's lab notebooks—written in Colorado Springs in 1899—that suggested a scientific platform for telepathy, or "supraliminal information transfer," as he called it.

"The mathematics that provided the underpinning for Tesla's work also provided the basis for understanding telepathy," and this in turn opened up "new frontiers for mind sciences," she said, providing "an extensive laundry list of the creative potentials of the human mind."

The more technical information she gave me about Tesla's technology, the more excited I became. It was a language I was accustomed to speaking from my work in the air force and NASA as well as at Disney. She showed me copies of some of his patents, which provided the technical basis for his inventions and technology.

In his autobiography, which I read on Valerie's recommendation, Tesla reported his uncanny ability to use "creative visualization." He would picture a particular apparatus, test-run the device, disassemble it, and check for proper action and wear—all through visualization. He was said to be capable of judging the dimensions of an object to a hundredth of an inch and to perform difficult computations in his head without the use of a slide rule or mathematical tables. When manufacturing his inventions, he worked with all the blueprints and specifications in his head, which has made some of his experiments difficult to duplicate. Tesla also reported having had a vision of some kind while taking a walk in a park one day. He claimed to have seen his "first rotating magnetic field," which gave him a key of sorts and led to some of his scientific accomplishments.

Valerie was convinced that not all of Tesla's knowledge originated on Earth. She thought that perhaps he—like she, the space kids, and others—had been selected to receive "subliminal transmissions" from sources of higher intelligence. After all, Tesla had been far ahead of everyone else at the time, and the visualization in the park he'd reported sounded to her a lot like telepathic transmissions or even remote viewing.

Had the same source of intelligence that had provided her with the space shuttle warning and other information been helping humans for a century or more? she wondered aloud. *Were technological transfers from other civilizations responsible for some of mankind's greatest scientific breakthroughs?*

Throughout the years, her own transmissions had often made mention of Tesla's work, in some cases listing his notes and patents for reference.

On one of my trips to Washington, D.C., to meet with Valerie at her Georgetown office, she presented me with a detailed proposal, "How to Build a Tesla Transmitter." Synthesizing the work of three of our primary technical contributors, she had written a draft document that was costed down to the penny.

I had seen and heard a good many fantastic claims since our first lunch more than a year ago. But here in my hand was a blueprint for introducing a clean, nondepletable energy system that had universal technological applications on a global scale. According to Valerie, a prototype could be built and tested in the southern California desert near Palm Springs for the grand sum of $64,400. The proposal went on to estimate that the initial pilot production plant would then require a $1.5 million investment.

Valerie suggested that we make an appearance before the Science and Astronautics Subcommittee of the House of Representatives, as well as the Atomic Energy Commission. She prepared a detailed campaign to educate the American people, our leaders, and the power companies about how this new technology might be integrated into the existing power grid. Lists of well-placed individuals representing corporate, civic, and private partnerships were laid out for my review.

Valerie's research on Tesla had been meticulous. Drawing on her training as an investigative reporter, she had pieced together from the work of this twentieth-century genius what she described as a "global puzzle," which was nothing short of a history of the evolution and suppression of free-energy technology on our planet.

With great enthusiasm, Valerie brought in more Tesla technology as possible projects for our researchers to explore—breakthrough inventions and proven scientific theories from half a century ago still not widely accepted or used. She was hopeful that acceptance would be found for this technology once it was tested and proved.

How far ahead of his time was Tesla? Forty years before Sputnik, he wrote an article that gave detailed and accurate descriptions of present-day guided missiles and rockets based on remote control. *How could he have known?*

Tesla's interest in aviation intrigued me. Documented through numerous patents were inventions considered impractical for his day and never developed, including a vertical-take-off aircraft that resembled in appearance the modern helicopter. And it was only during World War II that radar, a concept first described by Tesla in 1917, was developed.

As for space, Tesla's theories would contribute to its exploration, even though he didn't live to see it. In 1921 he first examined the possibility of a link being established with the planets of our solar system by means of ultrashort waves sent into space. Using his principle, the first ultrashort waves were sent by radar to the Moon and the Sun in 1946. When the beams bounced back to Earth, they brought with them the first scientific data on the exact distances these celestial bodies were from Earth.

If I had to choose a single person in history with whom to sit down for an hour-long chat, it would be Nikola Tesla, for the chance to hear about his inventions and future technological possibilities for mankind.

Tesla's inventions, which numbered eight hundred, replaced coal, eliminated the steam engine, and introduced modern-day electricity from industry to private homes. The areas in which he worked were far-reaching.

He was using fluorescent lights in his laboratory forty years before the industry "invented" them, and he demonstrated the principles used in microwave ovens decades before they became an integral part of our society.

His patents for an AC electric motor, purchased by Westinghouse in 1888, gave that company leadership in the field. From Tesla's early model evolved a wide range of motors in use today, from 1/10 horsepower to giants of over 60,000 horsepower.

Applied in the practice of medicine, Tesla's electronic advances have brought millions of people back to life from heart attacks with defibrillator units. Only recently has it become known that he made successful photographs of inner parts of the body by means of "very specific" waves some three years before a German researcher is credited with developing the use of X rays.

Another of his breakthroughs: designing and building prototypes of a fuel-burning rotary engine that was based on his earlier design of a rotary pump. Recent tests conducted on the Tesla bladeless disk turbine indicate that if built using today's high-temperature–resistant materials, it would rank as the world's most efficient gas engine.

Why some Tesla technology has long been overlooked is not hard to figure out. During his lifetime, Tesla found himself up against some powerful forces: power companies, oil companies, automobile manufacturers, and financial institutions.

If we were to move ahead with Tesla's technology, I wondered, would Valerie and I end up facing some of the same institutional opposition?

Tesla died a penniless recluse in 1943 at the age of eighty-seven in the New York hotel room that had served as his laboratory after a mysterious fire destroyed his research facilities, technical papers, and most-prized prototypes. Three months after his death, the U.S. Supreme Court nullified Marconi's patents on radio transmission, declaring that Tesla's patents predated them. Marconi, the court declared, had simply copied his former employer's work. But go into any middle-school science class in America and ask who invented the radio, and you'll hear the name Marconi, not Tesla.

It was the scientific community's loss to bury Tesla's achievements and not include or emphasize the study of his principles in elementary or secondary education. He is seldom mentioned with the "greats of electricity"—Faraday, Hertz, Marconi, Bell, Edison. Possibly because his life did not culminate in wealth and acclaim, Nikola Tesla has unfairly slipped from our national memory.

Tesla's wireless power transmission was reported at the time as an experimental reality, but the "secrets" of implementation were said to have died with the inventor, who wrote few things down. This is not so, for modern-day research and study of Tesla's patents—some by researchers we had lined up to work with—have shown that sufficient data are available to prove his theory a reality.

Valerie showed me photographs of Tesla's central power plant, transmitting tower, laboratory, and various inventions. She also produced several pieces of correspondence signed by Tesla, which alluded to his often futile attempts to generate interest in his discoveries.

The feasibility of wireless power transmission was proved by Tesla between July 1899 and February 1900 in Colorado Springs. His main areas of wireless research: sending Hertzian wave signals of very low frequency, and transmitting power based on creating a conductive path between the ionosphere and the Earth.

Tesla found that the Earth's surface could be used as a basic long-haul circuit for very low frequencies and that electrical energy could be transmitted worldwide from Earth—by going *through* the ground and using the ionosphere as a return path. This he accomplished in Colorado with little loss of energy. (Today's hardwired power grid loses up to 10 percent for every sixty miles that energy is transmitted.) "It is difficult to form an adequate idea of the marvelous power of this . . . [with] which the globe will be transformed," wrote Tesla of his experiments.

Today we know the ionosphere as an electrically conducting spherical shell of ions and free electrons that surrounds the Earth in the upper atmosphere—between fifty to two hundred miles high. And we know how important it is in radio communications, serving as a reflector of radio waves over a range of frequencies that permits transmission beyond lines of sight and around the Earth by successive reflections. In Tesla's time, however, little was known about this phenomenon, and much of the apparatus and components that would be needed to carry out this type of transmission had not yet been designed and built.

The possibilities for the wireless transmission of power are intriguing. Natural oil or gas deposits in a remote region of the world could be converted into electrical energy at the wellhead, or huge hydroelectric power projects could be developed in inaccessible locations and the electric power could then be transmitted—or beamed—to the user's location, potentially anywhere in the world, without the need for a network of high-voltage long-haul transmission wires.

Unfortunately, breakthrough technology sometimes has a dark side.

Valerie was certain that these electrical signals could produce dramatic effects on our ionosphere and showed me some of Tesla's original notes that alluded to specific physiological effects—good and bad—that these oscillations could produce.

Among the U.S. military and intelligence community, there were alarming reports that the USSR was engaged in large-scale efforts to develop a form of wireless radio transmission capable of affecting the behavior patterns of entire populations. That there was this type of activity seemed without question, as the Canadian Department of Communications recorded high-power low-frequency transmissions coming from the Soviet Union. Independent researchers verified similar transmissions originating from various sites in the Soviet Union. In intelligence circles, the Soviet signals became known as "Woodpecker" because they had a distinctive tap-tap-tap sound over the airwaves.

One of the most brilliant among the researchers we had lined up for the Advanced Technology Group was Dowd Hansen, an electronics genius who had an integral understanding of Tesla's technologies and could replicate some.

With a mass of unkempt hair, a reddish complexion, and a permanently disheveled look, Hansen had a definite look of Einstein about him. He was director of a research laboratory in California and had a bachelor's in engineering and a doctorate in science. For ten years he had been investigating the psychophysiological sensitivity of animals and humans to magnetic and electrical fields in extreme-low-frequency (ELF) ranges corresponding to brain waves. He had developed sensitive and reliable monitors, including an extremely low-noise, highly filtered amplifier capable of recording and displaying naturally occurring magnetic oscillations in our atmosphere.

I loved sitting down with Dowd and finding out the latest thing going on in his prolific mind. As brilliant as he was, he could be just as ornery to deal with personally, and as a result had few friends. I think he looked forward to our talks.

One day he was explaining to me about the "ionospheric wave guide" and how wave signals of very low frequency could be transmitted around the world without benefit of wires or antennas—just as Tesla envisioned. As a result of his low-frequency research, Hansen knew it worked.

I listened, then asked if such a system could be made secure.

He said it certainly could.

"Can you build a prototype for a demo?" I asked.

"Sure. Why?"

Having heard about the need for improved communications with our submerged nuclear submarines around the world, I suggested he arrange to give the U.S. Navy a demonstration of the Tesla-influenced communications system. It took more than a little nudging before he agreed, but the navy soon gave him the job of helping to design and implement the system. Today Hansen's work serves as the backbone of a very efficient worldwide communication system with our subs.

And the flip side . . .

Dowd Hansen was called in during the late 1970s to investigate widespread cattle deaths in Oregon. After some work in the field and follow-up in his laboratory, it was his opinion that the cattle had been killed by adverse ELF radio frequencies. The culprits? According to Hansen: *the Russians and their low-frequency transmissions.*

Dowd told me that researchers had for a few years been looking at the possibilities that man is a "biocosmic transducer"—not only a transmitter but a receiver too—and that somehow our brain waves can lock on and modulate with the Earth's electromagnetic field (which Tesla referred to as the Universal Magnetic Field, or UMF). Dowd speculated that this might be the medium for extra-sensory perception (ESP). According to Dowd, good work had been done on the subject at UCLA in the brain research group, and various government grants were sponsoring research into the effects of very-high- and very-low-frequency fields. He said published papers showed that these signals could influence the brain waves of cats and monkeys.

"What about humans?" I asked.

"It's a very sensitive subject among researchers," he said, "but yes, humans are affected. But as yet we don't know why."

From his own work, he knew at what levels humans were affected. Anything above 11 hertz (cycles per second) produces a range of general agitation or uneasiness. High-voltage power lines throw off 50 to 60 Hz, and there has long been concern about what such high frequencies can do to people living nearby. Frequencies under 7 Hz result in general feelings of relaxation, known as the "alpha state." The most beneficial frequency on Earth is believed to be 6.8 Hz. Interestingly, the Cheops Pyramid, built by the Egyptians three thousand years before Christ and about five thousand years before the invention of electricity, has a constant 6.8 Hz signal running through it. Various experts have measured and studied it but still do not know where the frequency comes from or why it is present in such an ancient structure.

A real mad-scientist type, Dowd had invented an electromagnetic pendant that gave off alpha frequencies and wore it around his neck. He claimed it not only increased his own sense of well-being but also beneficially affected others around him. "Gordo, it attracts women like magnets," he told me with a sly grin.

"We are electrical people," he explained. "We can be disoriented by electrical impulses." Our hearts could be made to start, *or stop,* by electricity, he said. "Electromagnetic radiation may be the most harmful pollutant in our society. There is strong evidence that cancer and other diseases can be triggered by electromagnetic waves."

Dowd had worked with Dr. Puharich in Ossining, designing receiving equipment to measure ELF waves. Their experiments showed that low-frequency signals could even penetrate the copper-lined walls of a Faraday cage. What protection did any of us have from the frequencies being used by the Russians? Several of them were thought to cause depression in humans.

Dowd said that when the Russians first started transmitting in 1976, they had emitted an 11 Hz signal through the Earth—just as Tesla had sent signals through the ground in Colorado some seventy-five years earlier. This ELF wave was so powerful that it upset radio communications around the world, and many nations lodged protests. The U.S. Air Force eventually identified five different frequencies the Russians were emitting in a wild ELF cocktail. "They never sent out alpha waves," Dowd explained. "Nothing in the six or seven hertz range that would be beneficial. They had more sinister things in mind."

ELFs will penetrate anything and everything, Dowd said. "Nothing stops or weakens these signals. At the right frequencies and durations, whole populations could be controlled by ELFs."

"Once the killing range is perfected and it's operational, this weapon will render nukes obsolete," he said, doom in his voice. "It can kill almost immediately with powerful adverse frequencies. Men, women, and children could be wiped out indiscriminately without destroying buildings, bridges, or cities. The conquering army walks in and takes over a completely undamaged infrastructure."

Through the years I had heard many scenarios about how future wars might one day be fought, and most of them were very frightening. But the scenario that Dowd Hansen presented was the coldest and most horrific I had ever considered.

Hot coffee on the stove, meals on the table, engines purring—

—and every living soul dead from adverse radio signals transmitted from half a world away.

Two decades after this conversation with Dowd Hansen, I know nothing about the present status of research—if any—by our country or other nations on this potential "ultimate weapon."

However, Hansen has since told me he believes that the former USSR came very close to having such an operational weapon.

Prior to his death in 1943, as World War II was peaking, Tesla claimed to have developed a "death-ray" weapon. Nothing ever came of it—some people speculated the inventor might have stopped work on it because of its destructive powers. "Peace can only come as a natural consequence of universal enlightenment," Tesla once said.

Upon his death, his hotel room was immediately searched by FBI agents looking for the design of the "death-ray" machine. Supposedly, his scant notes were all turned over to his native land, where a Tesla museum was built in Belgrade.

The push behind our 1980s "Star Wars" missile defense system was widespread fear that the Soviets had begun deployment of weapons based upon Tesla's high-energy principles. Reports of mysterious "blinding" of U.S. surveillance satellites, and the evidence of radio-signal interference gave credence to those concerns.

I can only assume that if low-frequency weapons research is under way in the United States or elsewhere, it must be cloaked in the highest level of secrecy imaginable.

I pray it will not be Nikola Tesla's final legacy to mankind.

RESERVATION FOR AN
ALIEN SAUCER RIDE

I had never met anyone who claimed to have ridden in an alien saucer.

Dan Fry looked every inch what he was: rocket scientist, researcher, electronics engineer. Favoring the tweed-jacket-with-leather-elbow-patches look, he was middle-aged with a receding hairline, mild-mannered, intelligent, quiet. A devout Christian, he practiced his beliefs. I don't suppose he'd ever said a naughty word in his life. He held a Ph.D. from St. Andrews College in London, England, where his doctoral thesis, "Steps to the Stars," prophesied much of the manned and unmanned space explorations of the 1960s and 1970s.

Valerie Ransone brought Dan Fry to my Disney office in 1979. Beforehand she assured me I would hit it off with this fellow because we had a good deal in common, most notably our "penchant for adventure" and "interest in a good UFO story." She added, "You've both experienced the real thing, Gordo."

When they walked in—Dan in a plaid shirt and smiling affably—he came over and gave me a firm handshake. She was right that we had crossed some of the same paths, and I liked him right away.

An early propulsion expert, he had developed a number of parts for the guidance system of the Atlas missile. He had worked for Aerojet General and been assigned to White Sands Proving Ground near Las Cruces, New Mexico, where he was in charge of instruments for missile control and guidance. He had also served as a consultant for the California Institute of Technology.

This wasn't the kind of guy you found on the next bar stool telling a tall tale about little green men from Mars, but he *did* have a fantastic story to tell about extraterrestrials. He related it in his unassuming manner, which suggested that he didn't particularly care if anyone believed him or not. In fact, he understood that many people wouldn't accept his story as "anything but fantasy." He still seemed somewhat amazed—thirty years later—by what had happened to him.

Dan Fry and I had seen our first UFO within a few months of each other.

On the evening of July 4, 1950, Dan had intended to go into Las Cruces with other scientists and engineers at the White Sands Proving Grounds, have dinner, and see some Fourth of July fireworks. Most of the Aerojet group left in a company car early in the afternoon, but Dan stayed behind to finish some work. He planned to take a bus into town later and meet up with them. But he missed the last bus and found himself stranded in a practically deserted camp with nothing much to do except sit in his room and read.

Around 8:30 P.M., he decided to take a walk. He stepped outside into a star-filled night and headed in the direction of the V–2 static test stand on which his Aerojet crew had been mounting a large new rocket motor for testing. The test stand was about a mile and a half from the camp.

About halfway to the test site he changed course, veering off onto a dirt road that led toward the base of the Organ Mountains. The road wasn't much more than a pair of wheel tracks in the desert, and he soon wound up among the sand dunes. The sun had set an hour earlier, but there was still enough light for a hike.

He scanned the clear sky, observing a group of bright stars just over the peaks of the mountains. As he did, one of the stars suddenly went out. He knew stars didn't just go out in a cloudless night sky and that something had to be eclipsing it.

If it was an airplane, it would have taken only a second or two before the plane passed by and the star came back into sight. But it didn't. He knew that weather balloons weren't released at night, and in any case a balloon would have risen rapidly and eclipsed the star only for a few seconds.

Then another star next to the first one went out, followed by two more.

Dan stopped in his tracks.

Whatever was cutting off the starlight was increasing rapidly in size. He waited, and then he saw it: a dark object approaching his position.

When it came closer, he understood why he hadn't seen it earlier—its color was nearly identical to that of the night sky.

His first inclination was to run, but from his long experience with rockets he knew it was foolish to run from an approaching object until you were certain of its trajectory—you might just as easily run into its path as away from it.

When the object was about a hundred feet away, he could make out a spheroid shape moving slowly toward him. Its reduced speed was reassuring. It did not look as if it was going to crash. Whatever was guiding it seemed to be in complete control. He watched as it glided in for a soft landing about seventy feet away.

The only noise was the crackling of the brush beneath the object, since the craft's operation had been silent. No propellers whipped the air, and no flash or roar of incandescent gases came spitting out of nozzles to produce thrust. The ship had coasted in, slowing to only a few miles per hour, and settled down without any sign of falling.

From his work with aircraft and missiles, Dan knew that the craft before him was more advanced than anything in the U.S. arsenal. His first thought was that it might be some secret technology belonging to the Russians, who he'd heard were ahead of us in the development of large rockets. But this wasn't a rocket.

Whatever this vehicle might be, it had operated efficiently and effortlessly in violation of the law of gravity. The builders of this machine had found the answers to a number of questions that had eluded our best physicists. For that reason, Dan had a strong feeling that the object before him had not been built in the Soviet Union—or anywhere on Earth.

The scientist in him took over, trying to figure out just how to handle the situation. He could return to base and report the strange craft, but it would probably take an hour to walk back, find someone in authority, and return with witnesses.

What if the object took off in the meantime? There would be nothing but a crumpled patch of brush left behind. What could be learned from that?

Deciding to investigate on his own, he approached the object slowly until he was within a few feet of it. It was about thirty feet in diameter at its widest point, which was about seven feet above the ground. The object's vertical dimension was fifteen to sixteen feet. Its curvature was such that, from below, the craft could appear to be saucer-shaped, when actually it was more like a soup bowl inverted over a saucer.

The dark blue color he had first observed was now gone, replaced by a silvery metallic finish. Walking around the craft, he was surprised to find no openings, hatches, or visible seams.

He touched the polished metal surface, carefully at first in case it was hot. He found it to be only a few degrees above the air temperature and incredibly smooth.

"Kind of like running your finger over a large pearl that had been covered with a thin film of soap," he told me. "Never felt anything quite like it."

Intrigued, he stroked the ship's surface with the palm of his hand, feeling a slight but definite tingling in the tips of his fingers and hand that was not unpleasant.

Then a voice came out of nowhere: *Better not touch the hull.*

He leaped back several feet. Losing his balance, he sprawled in the sand.

"I heard something like a chuckle," Dan said, smiling at the memory. "Then it said something about my taking it easy because I was *among friends.*"

After Dan got up, the voice told him that the reason for not touching the hull had to do with the fact that it was protected by *a field that repels all other matter*. The field was used for *lowering air friction when traveling at high speeds* through an atmosphere.

Perhaps you noticed that the surface seemed smooth and slippery. That is because your flesh did not actually touch the metal but was held a short distance from the surface by the repulsion of the field. There is a reason for this. Contact exposure to the metal causes human skin to produce what you call "antibodies" in the bloodstream. For some reason we don't yet fully understand, these antibodies are absorbed by the liver, whose function they attack, causing the liver to become enlarged and congested. It would have taken several months, but death would have been certain.

Dan was glad he hadn't poked too much into the soapy surface.

There were many things for the curious scientist to ask about, and he did. He soon had a feeling that the voice he was hearing was not coming to his ears as sound waves but rather was originating directly in his brain. They seemed to be communicating back and forth through some type of telepathy.

When he asked about this, the voice acknowledged that was indeed the case.

The saucer was a remote-controlled vehicle, he was told. It was being commanded from a *central unit, or what you would call the "mother ship,"* which was orbiting the Earth four thousand miles up.

"Why are you telling this to me and not someone else?" Dan asked.

Don't underestimate our ability to select the ones to whom we wish to speak.

Dan was told not only that he was a good "transmitter" but also a gifted "receiver." It was something that other humans were capable of doing, the knowing voice explained, but that a limited number of people on Earth had actually experienced.

"If I hadn't seen it with my own eyes and touched it for myself," Dan told me, "I might well have concluded that I was hallucinating. My science background had conditioned me to be prepared for the possibility of almost anything, but never in my wildest dreams did I think I would one day be standing in the desert talking to a visitor from another world via thought waves."

Dan shook his head in amazement and shrugged nonchalantly. His demeanor was take-it-or-leave-it, believe me or don't, your choice.

I had seen my share of wide-eyed UFO fanatics and lunatics. Dr. Dan Fry was *not* in that category. I found him totally credible.

Dan said the entity that spoke to him that night claimed to be *from a place far removed* from our solar system. *Previous expeditions to your planet by our ancestors over a period of many centuries met with almost total failure.*

"Total failure?" I said.

"He explained that one of their principal purposes for visiting Earth had been to determine our ability to adapt our minds quickly and calmly to conceptions that are completely foreign to our customary modes of thought."

"You mean, were we ready to hear that there was an extraterrestrial intelligence greater than our own?" I asked.

"Exactly. Apparently we weren't."

Then Dan got to the best part: he was invited to take a short flight.

Even though the remote-control saucer was only a "small cargo carrier," it did have a small passenger compartment containing four seats. A bottom portion of the hull opened up to create an opening some five feet in height and three feet wide, which he was able to enter standing up.

He sat down in one of the seats, and was surprised to find no straps.

"I was told I wouldn't feel any sensations or ill effects from the flight. And I didn't feel any acceleration at all or have the slightest sense of motion. It was explained that the force accelerating the vehicle was identical in nature to a gravitational field. It not only acted upon every atom of the vehicle itself but also equally upon every atom of mass that was within it, including me. I remember him saying, *The only limit of acceleration is the limit of available force*."

Even though the descriptions of space flight that Dan said his hosts provided may still not fit today's scientific knowledge, this tells me we have much to learn when it comes to traveling among the stars.

"The next thing I knew, we were ten miles high and I was looking down at the lights of El Paso. I could even distinguish the thin dark line of the Rio Grande River separating El Paso from its Mexican twin, Ciudad Juarez."

He could *see* all these sights even though the saucer had no windows. "A portion of the hull became transparent," he explained, "and I looked out as if I was seeing through a plate glass window at Macy's."

Dan said the encounter left him feeling depressed about his work and life, feeling they had lost meaning and significance. "A few hours before it happened," he said, "I'd been a self-satisfied engineer working on testing one of the largest rocket motors ever built by man. After the ride I took, I knew that the motor we had placed on the test stand was pitifully inefficient. It wouldn't take us very far into space."

Only when Dan finished did Valerie speak up. She explained that there had been a reason for her to introduce us. She had done so on instructions from a recent interdimensional transmission.

"Dan's already had one ride, but he's down for another. With you, Gordo," she added, flashing a smile. "You are to go together within the year."

"A *ride*?" I asked, somewhat perplexed.

"In an extraterrestrial spacecraft, Gordon," she said. "Are you interested?"

I had never been able to turn down a flight in anything, and I sure wasn't going to start now.

Dan was ready too. At the conclusion of the White Sands episode, he'd been told that he would be contacted again. He was thrilled about the chance to take another spin.

Valerie had received mental "pictures" of where we were to go. On the appointed day, she was to drive us to a spot in the desert outside of Yuma, near the California-Arizona border. She was given a list of items we should bring. In addition to a compass, they wanted me to bring a camera with infrared film. They said I would be able to take pictures that I could make public. It was to be, it seemed, a kind of coming-out party.

So had the ETs decided Earth was now ready?

I'd be less than honest if I said my wife was crazy about Valerie, who seemed to be the kind of woman who got along better with men than other women. Suzan was down-to-earth. There was nothing gullible about her either. She listened closely to what people said and watched their body language. In short, she was a darn good judge of people.

Her take on Valerie Ransone? Suzan wasn't sure. "Just be careful, Gordon," she had advised.

Now I went home and told Suzan about the invitation for a ride in an extraterrestrial spacecraft. She didn't tell me I was insane for wanting to take that drive into the Arizona desert with Valerie Ransone and Dan Fry. She heard me out, mulled it over for a day or two, then brought up the subject before bedtime one night.

"You're a test pilot," she said. "If there's any chance this is really going to happen, you should do it."

I loaded my camera, gathered my things, and waited—with the same anticipation that I'd waited for other space rides.

WINDOWS IN TIME
AND SPACE

For a while, the promise of new technologies kept the dream alive. The scope of the work being done by the numerous researchers prepared to join the Advanced Technology Group was the stuff of science fiction.

- Experiments in England with a *self-generating electromagnetic propulsion system* for a small unmanned saucer-shaped model; the system needed no fuel supply of its own but simply gathered electromagnetic particles in the atmosphere as it flew. From reports I reviewed and witnesses I spoke to who observed a test flight of the electromagnetic propulsion system, the model flat-out flew. But there was a problem with throttle control: it kept going and going, and as far as anyone knew, was still going—straight out into space with an unending supply of fuel.

  Electromagnetic propulsion will solve the problem of having to carry huge amounts of fuel on long voyages to distant planets. Such a system was envisioned by Nikola

Tesla decades ago. Since then, we have measured these electromagnetic particles—called neutrinos—which are prevalent in space. They've been found to travel at several times the speed of light. According to Einstein's original laws, the speed of light—186,000 miles per hour—was the highest speed any matter could attain. That speed limit, in the vastness of space, makes travel between even relatively close celestial neighbors impractical considering our life span. Four hundred light years in space is roughly equivalent to the blink of an eye. But just before he died, Einstein was in the process of rewriting his own laws, using time as a variable in space, not a fixed concept.

I'm convinced that electromagnetic energy will be the basis for future interstellar travel, and that man one day will travel in excess of the speed of light. Many experts today scoff at that "revolutionary" notion, but we had similar experts telling us forty years ago that the speed of sound (700 miles per hour) was the ultimate "sound barrier," and anyone who exceeded it would disintegrate along with his aircraft.

- A *medical device called a Multiple Wave Oscillator,* first developed years earlier by a French researcher and improved with solid-state electronics by one of our technical contributors. It used a miniature Tesla system, consisting of two copper coils. When activated, a magnetic field was established between the two coils, and the frequencies could be adjusted for the electricity that flowed from one coil to the other. The result was multiple waves of inherently "good frequencies" to help heal various parts of the body. I've used it myself successfully on various small injuries, placing a sore knee or strained muscle inside the magnetic field for a few minutes at a time. We put another willing volunteer on the machine: a Los Angeles Rams running back who had injured his Achilles tendon and was told by doctors he would be out for the rest of the year. After several treatments, he was back playing in a couple of weeks.

Two weeks before Dan Fry and I were to take our extraterrestrial saucer ride, Valerie Ransone showed up in my office looking as if she had been dragged behind a tractor. She was terribly depressed.

She would say only that there had been some strong disagreements among "certain parties" about Dan and I taking a ride, and that the ride was off.

"Someone teed off?" I asked.

She nodded.

"Will it be rescheduled?" After all, I *was* accustomed to launch delays.

"Now's not a good time to talk about it," she said weakly.

I was very disappointed. Dan and I had both been meticulous in our preparation to document the flight, as instructed, and as the date approached, my anticipation had grown measurably. Unfortunately, there was no flight director to whom I could now appeal the decision to scrub the mission.

I would later learn from Valerie that she had been "directed" to go out into the desert where we were to meet our ride two weeks hence, and became disoriented and confused by an array of "scrambled" transmissions from different entities that were in disagreement about the ride-along invitation. "It was like the good guys against the bad guys and I was the Queen of Battle," she said. The sun went down and she had a difficult trek back to where she'd left her rental car. When she got there, the car was gone; stolen, it would turn out. Out in the middle of nowhere, she ended up hitchhiking to Tucson. She took a ride from "a big rowdy guy in a big Cadillac," she told me, laughing. "He wanted to know what a 'little thing' like me was doing out in the desert at night all by her lonesome. What am I going to tell him? That I had an aborted UFO experience?" The whole incident she described as the "worst misfire" she had ever experienced since she started receiving transmissions.

Valerie and I continued our efforts with the Advanced Technology Group until late 1980, when it became apparent that our dream would die from lack of funding. Those sources of venture capital we did find would only provide seed money in exchange for complete control of the new company, which was unacceptable.

We had gone everywhere looking for backing—banks, private individuals and foundations, giant aerospace firms, even Congress. Valerie and I gave anyone and everyone who would listen joint briefings, which we came to call our "dog and pony show." At the time, I was so completely sold that I never thought the right people would say anything other than "This is great technology. Here's some money. Go do it." But for the most part that didn't happen. On some occasions we had good meetings with important people only to have their promises wither. While I'm sure some thought we were wackos, others did such quick flip-flops that we began to wonder if there was competition in government and/or private industry for the types of technologies we were planning to take to market.

As I've said, this country doesn't have a great record when it comes to encouraging and accepting revolutionary new technologies. This is because they can change not only the way we live and work but also much of the power structure upon which our economy is based. For instance, would the big oil companies sit idly by while a start-up firm attempted to bring to market a free-energy device? Would they be any less inclined to thwart such an advance, which could provide electricity to the poorest regions of the world, than they had been during Nikola Tesla's time? I am not a conspiracy buff, but these are questions that should be asked.

In early 1981 Valerie and I were to speak at the First Global Conference on the Future in Toronto, Canada. We were scheduled to make two presentations: "Social and Political Ramifications of Extraterrestrial Contact" and "Tesla Vision Realized: Plan for Free Worldwide Energy Transmission."

Something came up at the last minute, requiring me to cancel. Valerie went before 850 people in a ballroom, feeling deep down, I would learn nearly twenty years later, that I was afraid to put "my butt on the line" in front of a large crowd about such controversial subject matter.

After that, Valerie and I drifted apart, equally disillusioned by our failed efforts to launch the Advanced Technology Group and provide a supportive and nurturing atmosphere for all those brilliant scientists and researchers we had lined up.

I later heard from someone who knew her that she had moved to Hawaii and died there. No sooner did I repeat that rumor than a woman who knew Valerie well and had herself been involved in the Ossining experiments called and berated me for spreading a falsehood. "Valerie is very much alive," she said.

"Where is she?" I asked.

I was told that Valerie didn't want to be found.

But I did find Valerie in early 1999, living in a major city in the southwestern United States, where she teaches high school. We recently got together and reminisced.

"We were so innocent and trusting when we went on the road with our dog and pony show," she said. "We never really knew the kind of resistance we were up against. It was one minute to twelve when things fell apart for us. The world would have been a different and better place now had we succeeded. As it is, I'm afraid we're not leaving a very clean litter box for the next generation."

Valerie still claims to receive transmissions, which she finds both exhilarating and exhausting. Her purpose is unchanged, she told me: to inform, educate, and assist in laying the groundwork for extraterrestrial "technology transfers" to help solve the myriad problems facing mankind and Earth. She says she has no personal need to be a martyr. "I'd like a small house on the beach," she says, "and my own school so I can work with kids who have extra creative capabilities and nowhere to use them."

Today, I am unwilling to bet my life that the source of the astute technological assistance and advice Valerie Ransone provided from time to time originated from an extraterrestrial source. The test pilot skeptic in me could never be 100 percent convinced about her astral connections until I saw her climb aboard an alien flying saucer and take off. But until that happens, and based on what I saw with my own eyes and was able to check out, I am at least *80 percent certain* that she is, as she claims, able to tap into a source of higher intelligence not of this planet.

For me, it always goes back to the space shuttle transmission. If the information about that very technical and potentially catastrophic design flaw hadn't come from such an intelligence source, then how did she know the flaw existed when I didn't, Ben James didn't, and none of the other highly trained NASA engineers did either, until she brought it to our collective attention?

How *did* she know?

EPILOGUE

I still believe we need to send a manned mission to Mars.

An unmanned vehicle, no matter how sophisticated, is only as good as the designer who programs it, telling it what to look for. Anything beyond that, anything *unexpected*, and it doesn't know what to look for, or even what it's looking at.

At NASA this apropos cartoon made the rounds: A little unmanned vehicle is shown having landed on an unknown planet. It has analyzed the soil and temperature and is sending back a signal to Earth that there are no signs of life. Meanwhile, lurking behind the little robot-controlled probe is a hulking hairy monster with a big club.

We sent probes to the Moon before we landed men there and got some valuable information about what to expect. But the scientific return on the manned missions was computed to be more than a million to one over the unmanned vehicles.

We will get a lot from the upcoming international space station. What will be most valuable is that such a large number of nations will be working together in space. It will be a testing ground in many ways. I am not sure, however, that the public's imagination will be reignited by the space station, which some news commentators have already referred to as an "orbiting Motel 6."

I was delighted to see the return to space of my friend and former Mercury 7 colleague John Glenn. What most people don't realize is that NASA has detailed medical records on John and the rest of us. We have all taken thorough physical exams every year since our space flights. John's going back up was a unique opportunity to see what an older cardiovascular system would do under almost identical conditions with those it faced decades earlier. The comparative data must have been fascinating.

John did great in every way, and medically he had no adverse effects at all. His flight went far to prove that physically, the average person can handle space travel these days. We've come that far from when astronauts were selected from groups of qualified military test pilots, to better withstand the rigors of space flight.

John's return to space did something else: it captured the public's attention again for space exploration. The scenes of people jamming Times Square to see live TV coverage of the launch, and thousands lined up on beaches in Florida for the liftoff, reminded me of those days, which don't seem so long ago, when there was no bigger story.

Asked by the news media what I thought of John's return to space at the age of seventy-seven, I said, "I think it's fine as long as I get to go to Mars when I'm seventy-seven."

What will keep the public's interest will be when we get back to the real nuts-and-bolts of exploring space and finding out what's really *out there*. To do that, we need to get back to the Moon, and go to Mars, and be looking down the road at undertaking missions to other planets—and expeditions like mining asteroids, which could prove to be so lucrative that it might even be done commercially.

My old friend and retired CBS newscaster Walter Cronkite, who along with ABC's Jules Bergman and NBC's Roy Neal did such a splendid job of covering the early manned space flights, recently concluded that 1969 is the one date that will be best remembered from our century. "Five centuries from now," Walter said, "I believe the twentieth-century date that will be remembered is the year the human race first journeyed from the Earth to the stars." He went on to suggest that one day in the distant future, Wernher von Braun might be remembered as the "new Columbus."

The financial benefits of the space program have been well chronicled. It has been estimated that the technology spin-offs have yielded a twenty-to–one return on the investment. Every bit as important: the rapid explosion of the high-technology industry, a direct result of the massive investments made in cutting-edge research and development for the space program.

Recently I made a trip to Houston and visited the Johnson Space Center, paying the entrance fee to the museum like the many thousands of tourists who go there each year to see a piece of this country's space history. Both of my spacecraft, *Mercury* and *Gemini,* are on display there, hanging from a star-filled ceiling not far from the one spacecraft I always wanted to fly and never did: *Apollo.*

Outside in the bright sun, I squinted up at the largest lawn ornament in the world: a $225 million Saturn V rocket lying on its side, withering in the hot sun like a beached whale. It had been built to take *Apollo 18* to the Moon in the early 1970s, but when the United States ran out of money and inspiration, it was ordered to stand down. I believe it's the only Saturn V still around, and even in such an ignoble position, it's impressive.

Four hundred feet long from its needlelike nose to four stubby fins, designed by a very good friend of mine on a scale more commonly used for huge public works projects—but this baby was built to go to the stars.

I walked back to my car, thinking, *What a terrible waste.*

•   •   •

I left Disney in 1980 and became a consultant on special projects with high-technology companies. I also make presentations to management on technology and industry for the future.

I'm still involved in aviation, and get cranky if I don't fly as often as I think I should. Since 1989 I've been a partner in an aeronautical design firm, Galaxy Aviation, in Van Nuys, California. One of the craft we are presently working on is called the Swift 2000, a saucer-shaped vertical-lift vehicle. It has four vertical-lift fans and should work very effectively. It will be able to go up, down, sideways—everything a helicopter can do, only much better. It will have tremendous lift capabilities and be able to carry heavy loads, such as cargo and water for fire fighting, in podlike compartments. I'm also working on a real speed demon: an unlimited pylon racer shaped like a saucer and powered by a big Chevy engine. I intend to enter it in the annual National Championship Air Races in Reno, and win.

I've never forgotten what I saw in the skies over Europe— or in Wendell Welling's barn.

I was asked recently what I would do if an alien spacecraft landed in my backyard and offered me a ride. "Would you go or stay, Gordo?"

Stepping aboard a ship from a distant world could result in something more than an afternoon joyride, I know. If there is interdimensional travel—and I believe there is—I could return to "Earth time" and find my loved ones and friends long deceased.

I love my wife and daughters, and I enjoy spending time with my friends and old flying buddies, and I would miss them all. But I'm still an astronaut at heart.

I'm going.

FAREWELL
TO A BUDDY

While riding with his wife, Nancy, and other friends, Charles "Pete" Conrad lost control of his 1996 Harley on a mountain road near Ojai, California, on July 8, 1999.

Pete died from his injuries five hours later in the hospital. He was sixty-nine.

After Pete and I flew together on *Gemini 5*, he commanded *Gemini 11*, a three-day mission with Richard Gordon. They conducted two EVAs and set a new altitude record of 850 miles. Pete commanded *Apollo 12*'s voyage to the Moon in 1969. He and Alan Bean walked on the dusty lunar surface collecting rocks and conducting experiments. Pete also commanded the first Skylab mission in May 1973.

Pete retired from NASA (and the U.S. Navy) not long after I did, in 1973, and worked at McDonnell Douglas for twenty years before retiring a second time in 1996.

Back in the 1960s, a bunch of us astronauts decided that whichever one of us died with the most toys would win. Pete had them all—motorcycles, boats, planes, fast cars.

We are now all in agreement.

Pete won.

INDEX

T 48572